YET MORE
EVERYDAY
SCIENCE MYSTERIES

STORIES FOR INQUIRY-BASED SCIENCE TEACHING

YET MORE EVERYDAY SCIENCE MYSTERIES

STORIES FOR INQUIRY-BASED SCIENCE TEACHING

Richard Konicek-Moran, EdD
Professor Emeritus
University of Massachusetts
Amherst

Botanical illustrations by
Kathleen Konicek-Moran

National Science Teachers Association

National Science Teachers Association

Claire Reinburg, Director
Jennifer Horak, Managing Editor
Andrew Cooke, Senior Editor
Judy Cusick, Senior Editor
Wendy Rubin, Associate Editor
Amy America, Book Acquisitions Coordinator

ART AND DESIGN
Will Thomas Jr., Director, cover and interior design
Additional illustrations by Kistofer Boban

PRINTING AND PRODUCTION
Catherine Lorrain, Director
Nguyet Tran, Assistant Production Manager

NATIONAL SCIENCE TEACHERS ASSOCIATION
Francis Q. Eberle, PhD, Executive Director
David Beacom, Publisher

LIBRARY OF CONGRESS CATALOGING-IN-PUBLICATION DATA
Konicek-Moran, Richard.
  Yet more everyday science mysteries: stories for inquiry-based science teaching / Richard Konicek-Moran; botanical illustrations by Kathleen Konicek-Moran.
      p. cm.
  Includes index.
  ISBN 978-1-936137-11-4
  1. Science—Methodology. 2. Problem solving. 3. Science—Study and teaching. 4. Science—Miscellanea. 5. Inquiry-based learning.
6. Detective and mystery stories. I. Title.
  Q175.K6635 2011
  372.35'044—dc22
                    2010052783
eISBN 978-1-936137-47-3

CONTENTS

The Stories and Background Materials for Teachers

ACKNOWLEDGMENTS

This book is dedicated to a brilliant educator, Distinguished Professor of Humanics Robert Barkman of Springfield College, who has continually encouraged and supported my work and has used the stories and techniques in workshops with Springfield, Massachusetts elementary and middle school teachers over the past six years.

I would also like to dedicate these stories and materials to the dedicated and talented teachers in the Springfield Public Schools in Springfield, Massachusetts. They have been my inspiration to produce materials that work with city as well as rural children.

I would like to thank the following teachers, educators, and administrators who have helped me by field-testing the stories and ideas contained in this book over many years. These dedicated educators have helped me with their encouragement and constructive criticism:

Richard Haller
Jo Ann Hurley
Lore Knaus
Ron St. Amand
Renee Lodi
Deanna Suomala
Louise Breton
Ruth Chappel
Theresa Williamson
Third-Grade Team at Burgess Elementary in Sturbridge, Massachusetts
Second-grade Team Burgess Elementary in Sturbridge, Massachusetts
Fifth-grade Team at Burgess Elementary in Sturbridge, Massachusetts
Teachers at Millbury, Massachusetts Elementary Schools
Teachers and children at Pottinger Elementary School, Springfield, Massachusetts
All the administrators and science specialists in the Springfield, Massachusetts public schools, who are too numerous to mention individually

My thanks also go out to all of the teachers and students in my graduate and undergraduate classes who wrote stories and tried them in their classes as well as using my stories in their classes.

I will always be in the debt of my advisor at Columbia University, the late Professor Willard Jacobson who made it possible for me to find my place in teacher education at the university level.

I also wish to thank Skip Snow, Jeff Kline, Jean and Rick Seavey and all of the biologists in the Everglades National Park with whom I have had the pleasure of working for the past ten years for helping me to remember how to be a scientist again. And to the members of the interpretation groups in the Everglades National Park, at Shark Valley and Pine Island who helped me to realize again that it possible to help someone to look without telling them what to see and to help me to realize how important it is to guide people toward making emotional connections with our world.

Christiana Admiral, Pine Island District Interpreter, Everglades National Park,

Everglades National Park, Rangers, Maria Thomson, Laurie Humphrey Leon Howell, Kirk Singer, Rudy Beotegui, Frankie Aranzamendi, and the others who keep the Park a place of inspiration and devotion to sustaining out wild heritages.

My sincere thanks goes to Claire Reinburg of NSTA who had the faith in my work to publish the original book and the second and third volumes and is now taking a chance on a fourth; and to Andrew Cooke, my editor, who helps me through the crucial steps. In addition I thank my lovely, brilliant and talented wife, Kathleen for her support, criticisms, illustrations and draft editing.

Finally I would like to dedicate these words to all of the children out there who love the world they live in and to the teachers and parents who help them to make sense of that world through the study of science.

Preface

Teaching and Interpreting Science

Over the past nine years my wife and I have had the privilege of being nature interpreters in the Everglades National Park. We were warned that interpretation was different from teaching. We were not supposed to be lecturing about the names of birds or plants but helping the visitors tune into the beauty and value of the park. In fact, we were told that our major goal was to help the visitors make "an emotional connection to the resource (the park), not to teach."

So, it would seem that teaching and interpreting are quite different entities. I'm not sure I agree, or perhaps it is that I hope that they become more like each other. Synonyms for *interpret* are "enlighten, elucidate, clarify, illuminate or shed light on." *Teaching* is also defined as "to enlighten and illuminate." Most dictionaries, it is true, do not include "to help make an emotional connection to…" in their definitions. But I think this may be a great idea, in science as well as in all other subjects.

Science is a construction created by humans to make sense of the world. Over the centuries, science has been invented, reinvented, and modified. It follows, or is supposed to follow, certain rules by which it operates. At first it was known as natural history or natural philosophy, and debate was its favorite mode of operation. Later Galileo opened the door to direct experimentation, and scientists such as Kepler, Brahe, Newton, and Darwin showed how the interpretation of data can lead to explanations and theories that allow us to predict, with fair accuracy, events and everyday occurrences, or to develop the technology to do tremendous things, such as to send humans to the Moon.

This book is based on everyday occurrences and the desire to understand and enjoy them. It is paramount that the teacher and student make "emotional connections" to the world they are trying to understand. An emotional connection to a flower or worm or insect may not be absolutely essential to knowing about it, but making an emotional connection to each and every critter on our planet and to its place in the ecosystem helps us see how we are involved, along with every *other* thing on our planet, as a fully participating part of the entire system.

One does not develop *values* out of knowledge alone. Values are what we do when no one is watching (for example, walking to the trash can to deposit litter even though no one is around to witness it). A value is apparent when we compost our vegetable matter or recycle our aluminum, glass, and paper or take our own bags to the market, even though there is no law to demand it. We do so because we have made an emotional connection to our planet. Thus I posit that we need to help our students make emotional connections to the enterprise of science and to understanding how our world works, as best we can interpret it.

Visitors to the Everglades National Park should be impressed, for example, by the realization that the plants they are viewing have survived six months of drought and six months of deluge (the Everglades is semitropical, with wet and dry seasons). They should understand that the plants have survived the two things that usually kill them, underwatering and overwatering. Most visitors know this about plants, and we build on this everyday understanding to motivate the groups to look for those attributes each plant has developed to adapt to this harsh climate. They begin to notice the waxy leaf covering, the shapes of the leaves, dormancy behavior, and other special features that mitigate the damage of too much water; however, these same plants also retain water during those times when there is little available. Visitors are awed when they meet baby barred owls that find this environment to their liking. Thus, the emotional connection is made. Hopefully, this leads to an understanding of the importance of protecting such an environment.

The students in your classroom can have the same experience. They should, whenever possible, make that emotional connection to the ocean, to a lever, to the condensation of water on a glass, to forces that affect their lives and their bodies, and certainly to the process of science itself. And I believe that I can safely say that without some emotional connection to topics in a curriculum, little will really be learned, remembered, and understood. This is why stories about children who live lives like theirs can help them make these connections.

Recently, a poem appeared on my desktop, which seemed to support some of the things that I have been advocating in my work in these volumes of everyday science mysteries. I ask you to remember the words and thoughts as you plan your teaching.

Leisure
by William Henry Davies (1921)

What is this life if, full of care,
We have no time to stand and stare.
No time to stand beneath the bough
And stare as long as sheep or cows.
No time to see, when woods we pass,
Where squirrels hide their nuts in grass.
No time to see, in broad daylight,
Streams full of stars, like skies at night.
No time to turn at Beauty's glance,
And watch her feet, how they can dance.
No time to wait till her mouth can
Enrich that smile her eyes began.
A poor life this if, full of care,
We have no time to stand and stare.

"everyday miracles"

I am often asked about the origin of these everyday science mysteries. The answer is that they are most often derived from my day-to-day experiences. Science is all around us as we go through our routines, but it often eludes us because, as the old saying goes, "The hidden we seek, the obvious we ignore."

I am fortunate to be surrounded by a rural natural environment. My daily routine is predictable. I arise, eat breakfast, and then walk with my wife through the woods for a mile or so to exercise our Australian shepherd and ourselves. Our dog acts as a wonderful model as she exhibits her awareness of every scent and sight that might have changed over the past 24 hours. Her nose is constantly sniffing the ground and air in search of the variety of clues well beyond our limited senses. As we walk, we look for our "miracle of the day." It may be a murder of crows harassing a barred owl or a red-tailed hawk flying over our heads with a squirrel in its talons. It might be a pair of wood ducks looking for a tree with a hole big enough for a nest or a patch of spring trillium or trout lilies. In the late summer, it could be a clump of ghostly Indian pipe and a rattlesnake plantain orchid in bloom or a hummingbird hovering near a flower fueling up for its long trip south. Today it is a gigantic, beautiful, mysterious, salmon-pink mushroom, never before noticed. Sounds from the road bring questions about how sound travels, and as we arrive home, we see crab apples, the worms in the compost pile, and the new greenhouse whose temperature fluctuations have plagued us all summer.

Textbooks are full of interesting information about the planets, space travel, plant reproduction, and animal behavior, but very little about how this information was developed. Our world is full of questions, many of which are investigable by children and adults. Our senses and mind are drawn to these questions, which stimulate the "I wonder…" section of our brains. We are intrigued by shadows, by the motion of the Sun and Moon during the daytime and the stars and planets at night. There are mysteries at every turn, if we keep our minds and eyes open to them.

I am amazed that so many years have passed without my noticing so many of the mysteries that surround me. Writing these books has had a stimulating effect upon the way I look at the world. I thank my wife, a botanist, artist, and gardener, for spiking my awareness of the plants that I glossed over for so many years. We can get so caught up in the glitz of newsworthy science that we are blind to the little things that crawl at our feet, or sway in the branches over our heads, or move through the sky in predictable and fascinating ways each and every day. One can wonder where the wonder went in our lives as we teachers get caught up in the search for better and better test scores. The stories spring forth by themselves when I can remember to see the world through childlike eyes. Perhaps, therein lies the secret to seeing these everyday science mysteries.

WRITING EVERYDAY SCIENCE MYSTERY STORIES

When I first started writing stories, I tried the idea out with a seminar of my graduate students. We selected science topics, wrote stories about phenomena, and added challenges by leaving the endings open, requiring the readers to engage in what we hoped would be actual inquiry to finish the story.

Things to think about as you write your story
Does your story…
1. address a single concept or conceptual scheme?
2. address a topic of interest to your target age group?
3. try to provide your audience with a problem they can solve through direct activity?
4. require the students to become actively involved—hands-on, minds-on?
5. have a really open-ended format?
6. provide enough information for the students to identify and attack the problem?
7. involve materials that are readily available to the students?
8. provide opportunities for students to discuss the story and come up with a plan for finding some answers?
9. make data collection and analysis of those data a necessity?
10. provide some way for you to assess what their current preconceptions are about the topic? (This can be implicit or explicit.)

We also added distracters—children's ideas and misconceptions—that were intended to double as formative assessment tools. Over the course of the semester we wrote many stories and tried them out with students in classrooms. The children enjoyed the stories, and we learned some important lessons on how to formulate the stories so that they provided the proper challenge.

For years afterward, I used the idea with my graduate and undergraduate students in the elementary science methods classes. In lieu of the usual lesson plan, my class requirements included a final assignment that asked them to write a story about a science phenomenon and include a follow-up paper that described how they would use the story to encourage inquiry learning in their classrooms. As I learned more about the concept, I was able to add techniques to my repertoire, which enhanced the quality of the stories and follow-up papers.

I found that teachers benefit from talking about their stories with other teachers and their instructor. They can gain valuable feedback before they launch into the final story. We organized small-group meetings of no more than five students to preview and discuss ideas. We also designed a checklist document, which helped clarify the basic ideas behind the concept of the "challenge story." (See box.)

As usual, practice makes for a better product and finally my students were producing stories that were useful for them and were acceptable to me as a form of assessment of their learning about improving their teaching of science as inquiry.

As the years went by, teachers began to ask me if my own stories, which I used for examples in class, were available for them to use. They encourged me to publish them. I hope that they will provide you with ideas and inspiration to develop more inquiry-oriented lessons in your classrooms. Perhaps you may be motivated to try writing your own stories for teaching those concepts you find most difficult to teach.

references

Davies, W. H. 2009. *Collected poems by William H. Davies (1921)*. Whitefish, MT: Kessinger.

Konicek-Moran, R. 2008. *Everyday science mysteries*. Arlington, VA: NSTA Press.

INTRODUCTION

CASE STUDIES ON HOW TO USE THE STORIES IN THE CLASSROOM

I would like to introduce you to one of the stories from the first volume of *Everyday Science Mysteries* (Konicek-Moran 2008) and then show how the story was used by two teachers, Teresa, a second-grade teacher, and Lore, a fifth-grade teacher. Then in the following chapters I will explain the philosophy and organization of the book before going to the stories and background material. Here is the story, "Where Are the Acorns?"

WHERE ARE THE ACORNS?

Cheeks looked out from her nest of leaves, high in the oak tree above the Anderson family's backyard. It was early morning and the fog lay like a cotton quilt on the valley. Cheeks stretched her beautiful gray, furry body and looked about the nest. She felt the warm August morning air, fluffed up her big gray bushy tail and shook it. Cheeks was named by the Andersons since she always seemed to have her cheeks full of acorns as she wandered and scurried about the yard.

"I have work to do today!" she thought and imagined the fat acorns to be gathered and stored for the coming of the cold times.

Now the tough part for Cheeks was not gathering the fruits of the oak trees. There were plenty of trees and more than enough acorns for all of the gray squirrels who lived about the yard. No, the problem was finding them later on when the air was cold and the white stuff might be covering the lawn. Cheeks had a very good smeller and could sometimes smell the acorns she had buried earlier. But not always. She needed a way to remember where she had dug the holes and buried the acorns. Cheeks also had a very small memory and the yard was very big. Remembering all of these holes she had dug was too much for her little brain.

The Sun had by now risen in the east and

Cheeks scurried down the tree to begin gathering and eating. She also had to make herself fat so that she would be warm and not hungry on long cold days and nights when there might be little to eat.

"What to do ... what to do?" she thought as she wiggled and waved her tail. Then she saw it! A dark patch on the lawn. It was where the Sun did not shine. It had a shape and two ends. One end started where the tree trunk met the ground. The other end was lying on the ground a little ways from the trunk. "I know," she thought. "I'll bury my acorn out here in the yard, at the end of the dark shape and in the cold times, I'll just come back here and dig it up!!! Brilliant Cheeks," she thought to herself and began to gather and dig.

On the next day she tried another dark shape and did the same thing. Then she ran about for weeks and gathered acorns to put in the ground. She was set for the cold times for sure!!

Months passed and the white stuff covered the ground and trees. Cheeks spent more time curled up in her home in the tree. Then one bright crisp morning, just as the Sun was lighting the sky, she looked down and saw the dark spots, brightly dark against the white ground. Suddenly she had a great appetite for a nice juicy acorn. "Oh yes," she thought. "It is time to get some of those acorns I buried at the tip of the dark shapes."

She scampered down the tree and raced across the yard to the tip of the dark shape. As she ran, she tossed little clumps of white stuff into the air and they floated back onto the ground. "I'm so smart," she thought to herself. "I know just where the acorns are." She did seem to feel that she was a bit closer to the edge of the woods than she remembered but her memory was small and she ignored the feelings. Then she reached the end of the dark shape and began to dig and dig and dig!

And she dug and she dug and she dug! Nothing!! "Maybe I buried them a bit deeper," she thought, a bit out of breath. So she dug deeper and deeper and still, nothing. She tried digging at the tip of another of the dark shapes and again found nothing. "But

I know I put them here," she cried. "Where could they be?" She was angry and confused. Did other squirrels dig them up? That was not fair. Did they just disappear? What about the dark shapes?

HOW TWO TEACHERS USED "WHERE ARE THE ACORNS?"

Teresa, a veteran second-grade teacher

Teresa usually begins the school year with a unit on fall and change. This year she looked at the National Science Education Standards (NSES) and decided that a unit on the sky and cyclic changes would be in order. Since shadows were something that the children often noticed and included in playground games (shadow tag), Teresa thought using the story of "Cheeks" the squirrel would be appropriate.

To begin, she felt that it was extremely important to know what the children already knew about the Sun and the shadows cast from objects. She wanted to know what kind of knowledge they shared with Cheeks and what kind of knowledge they had that the story's hero did not have. She arranged the children in a circle so that they could see one another and hear one another's comments. Teresa read the story to them, stopping along the way to see that they knew that Cheeks had made the decision on where to bury the acorns during the late summer and that the squirrel was looking for her buried food during the winter. She asked them to tell her what they thought they knew about the shadows that Cheeks had seen. She labeled a piece of chart paper, "Our best ideas so far." As they told her what they "knew," she recorded their statements in their own words:

"Shadows change every day."
"Shadows are longer in winter."
"Shadows are shorter in winter."
"Shadows get longer every day."
"Shadows get shorter every day."
"Shadows don't change at all."
"Shadows aren't out every day."
"Shadows move when you move."

She asked the students if it was okay to add a word or two to each of their statements so they could test them out. She turned their statements into questions and the list then looked like this:

"Do shadows change every day?"
"Are shadows longer in winter?"
"Are shadows shorter in winter?"
"Do shadows get longer every day?"
"Do shadows get shorter every day?"
"Do shadows change at all?"
"Are shadows out every day?"
"Do shadows move when you move?"

Teresa focused the class on the questions that could help solve Cheeks's dilemma. The children picked "Are shadows longer or shorter in the winter?" and "Do shadows change at all?" The children were asked to make predictions based on their experiences. Some said that the shadows would get longer as we moved toward winter and some predicted the opposite. Even though there was a question as to whether they would change at all, they agreed unanimously that there would probably be some change over time. If they could get data to support that there was change, that question would be removed from the chart.

Now the class had to find a way to answer their questions and test predictions. Teresa helped them talk about fair tests and asked them how they might go about answering the questions. They agreed almost at once that they should measure the shadow of a tree each day and write it down and should use the same tree and measure the shadow every day at the same time. They weren't sure why time was important except that they said they wanted to make sure everything was fair. Even though data about all of the questions would be useful, Teresa thought that at this stage, looking for more than one type of data might be overwhelming for her children.

Teresa checked the terrain outside and realized that the shadows of most trees might get so long during the winter months that they would touch one of the buildings and become difficult to measure. That could be a learning experience but at the same time it would frustrate the children to have their investigation ruined after months of work. She decided to try to convince the children to use an artificial "tree" that was small enough to avoid our concern. To her surprise, there was no objection to substituting an artificial tree since, "If we measured that same tree every day, it would still be fair." She made a tree out of a dowel that was about 15 cm tall and the children insisted that they glue a triangle on the top to make it look more like a tree.

The class went outside as a group and chose a spot where the Sun shone without obstruction and took a measurement. Teresa was concerned that her students were not yet adept at using rulers and tape measures so she had the children measure the length of the shadow from the base of the tree to its tip with a piece of yarn and then glued that yarn onto a wall chart above the date when the measurement was taken. The children were delighted with this.

For the first week, teams of three went out and took daily measurements. By the end of the week, Teresa noted that the day-to-day differences were so small that perhaps they should consider taking a measurement once a week. This worked much better, as the chart was less "busy" but still showed any important changes that might happen.

As the weeks progressed, it became evident that the shadow was indeed getting longer each week. Teresa talked with the students about what would make a shadow get longer and armed with flashlights, the children were able to make longer shadows of pencils by lowering the flashlight. The Sun must be getting lower too if this was the case, and this observation was added to the chart of questions. Later, Teresa wished that she had asked the children to keep individual science notebooks so that she could have been more aware of how each individual child was viewing the experiment.

The yarn chart showed the data clearly and the only question seemed to be, "How long will the shadow get?" Teresa revisited the Cheeks story and the children were able to point out that Cheeks's acorns were probably much closer to the tree than

the winter shadows indicated. Teresa went on with another unit on fall changes and each week added another piece of yarn to the chart. She was relieved that she could carry on two science units at once and still capture the children's interest about the investigation each week after the measurement. After winter break, there was great excitement when the shadow began getting shorter. The shortening actually began at winter solstice around December 21 but the children were on break until after New Years. Now, the questions became "Will it keep getting shorter? For how long?" Winter passed and spring came and finally the end of the school year was approaching. Each week, the measurements were taken and each week a discussion was held on the meaning of the data. The chart was full of yarn strips and the pattern was obvious. The fall of last year had produced longer and longer shadow measurements until the New Year and then the shadows had begun to get shorter. "How short will they get?" and "Will they get down to nothing?" questions were added to the chart. During the last week of school, they talked about their conclusions and the children were convinced that the Sun was lower and cast longer shadows during the fall to winter time and that after the new year, the Sun got higher in the sky and made the shadows shorter. They were also aware that the seasons were changing and that the higher Sun seemed to mean warmer weather and trees producing leaves. The students were ready to think about seasonal changes in the sky and relating them to seasonal cycles. At least Teresa thought they were.

On the final meeting day in June, she asked her students what they thought the shadows would look like next September. After a great deal of thinking, they agreed that since the shadows were getting so short, that by next September, they would be gone or so short that they would be hard to measure. Oh my!! The idea of a cycle had escaped them, and no wonder, since it hadn't really been discussed. The obvious extrapolation of the chart would indicate that the trend of shorter shadows would continue. Teresa knew that she would not have a chance to continue the investigation next September but she might talk to the third-grade team and see if they would at least carry it on for a few weeks so that the children could see the repeat of the previous September data. Then the students might be ready to think more about seasonal changes and certainly their experience would be useful in the upper grades where seasons and the reasons for seasons would become a curricular issue. Despite these shortcomings, it was a marvelous experience and the children were given a great opportunity to design an investigation and collect data to answer their questions about the squirrel story at a level appropriate to their development. Teresa felt that the children had an opportunity to carry out a long-term investigation, gather data, and come up with conclusions along the way about Cheek's dilemma. She felt also that the standard had been partially met or at least was in progress. She would talk with the third-grade team about that.

Lore (pronounced Laurie), a veteran fifth-grade teacher
In September while working in the school, I had gone to Lore's fifth-grade class for advice. I read students the Cheeks story and asked them at which grade they thought it would be most appropriate. They agreed that it would most likely fly best at second grade. It seemed, with their advice, that Teresa's decision to use it there was a good one.

However, about a week after Teresa began to use the story, I received a note from Lore, telling me that her students were asking her all sorts of questions about shadows, the Sun, and the seasons and aking if I could help. Despite their insistence that the story belonged in the second grade, the fifth graders were intrigued enough by the story to begin asking questions about shadows. We now had two classes interested in Cheeks's dilemma but at two different developmental levels. The fifth graders were asking questions about daily shadows, direction of shadows, and seasonal shadows, and they were asking, "Why is this happening?" Lore wanted to use an inquiry approach to help them find answers to their questions but needed

help. Even though the Cheeks story had opened the door to their curiosity, we agreed that perhaps a story about a pirate burying treasure in the same way Cheeks had buried acorns might be better suited to the fifth-grade interests in the future.

Lore looked at the NSES for her grade level and saw that they called for observing and describing the Sun's location and movements and studying natural objects in the sky and their patterns of movement. But the students' questions, we felt, should lead the investigations. Lore was intrigued by the 5E approach to inquiry (*engage, elaborate, explore, explain, and evaluate*) and because the students were already "engaged," she added the "elaborate" phase to find out what her students already knew. (The five Es will be defined in context as this vignette evolves.) So, Lore started her next class asking the students what they "knew" about the shadows that Cheeks used and what caused them. The students stated:

"Shadows are long in the morning, short at midday, and longer again in the afternoon."

"There is no shadow at noon because the Sun is directly overhead."

"Shadows are in the same place every day so we can tell time by them."

"Shadows are shorter in the summer than in the winter."

"You can put a stick in the ground and tell time by its shadow."

Just as Teresa had done, Lore changed these statements to questions, and they entered the "exploration" phase of the 5E inquiry method.

Luckily, Lore's room opened out onto a grassy area that was always open to the Sun. The students made boards that were 30 cm² and drilled holes in the middle and put a toothpick in the hole. They attached paper to the boards and drew shadow lines every half hour on the paper. They brought them in each afternoon and discussed their results. There were many discussions about whether or not it made a difference where they placed their boards from day to day.

They were gathering so much data that it was becoming cumbersome. One student suggested that they use overhead transparencies to record shadow data and then overlay them to see what kind of changes occurred. Everyone agreed that it was a great idea.

Lore introduced the class to the *Old Farmer's Almanac* and the tables of sunsets, sunrises, and lengths of days. This led to an exciting activity one day that involved math. Lore asked them to look at the sunrise time and sunset time on one given day and to calculate the length of the daytime Sun hours. Calculations went on for a good 10 minutes and Lore asked each group to demonstrate how they had calculated the time to the class. There must have been at least six different methods used and most of them came up with a common answer. The students were amazed that so many different methods could produce the same answer. They also agreed that several of the methods were more efficient than others and finally agreed that using a 24-hour clock method was the easiest. Lore was ecstatic that they had created so many methods and was convinced that their understanding of time was enhanced by this revelation.

This also showed that children are capable of metacognition—thinking about their thinking. Research (Metz 1995) tells us that elementary students are not astute at thinking about the way they reason but that they can learn to do so through practice and encouragement. Metacognition is important if students are to engage in inquiry. They need to understand how they process information and how they learn. In this particular instance, Lore had the children explain how they came to their solution for the length of day problem so that they could be more aware of how they went about solving the challenge. Students can also learn about their thinking processes from peers who are more likely to be at the same developmental level. Discussions in small groups or as an entire class can provide opportunities for the teacher to probe for more depth in student explanations. The teacher can ask the students who explain their technique to be more specific about how they used their thought processes: dead ends as well as successes. Students can also learn more about their metacognitive processes by writing in their notebooks about how they thought

through their problem and found a solution. Talking about their thinking or explaining their methods of problem solving in writing can lead to a better understanding of how they can use reasoning skills better in future situations.

I should mention here that Lore went on to teach other units in science while the students continued to gather their data. She would come back to the unit periodically for a day or two so the children could process their findings. After a few months, the students were ready to get some help in finding a model that explained their data. Lore gave them globes and clay so that they could place their observers at their latitude on the globe. They used flashlights to replicate their findings. Since all globes are automatically tilted at a 23.5-degree angle, it raised the question as to why globes were made that way. It was time for the "explanation" part of the lesson and Lore helped them to see how the tilt of the Earth could help them make sense of their experiences with the shadows and the Sun's apparent motion in the sky.

The students made posters explaining how the seasons could be explained by the tilt of the Earth and the Earth's revolution around the Sun each year. They had "evaluated" their understanding and "extended" it beyond their experience. It was, Lore agreed, a very successful "6E" experience. It had included the engage, elaborate, explore, explain, and evaluate phases, and the added extend phase.

references

Konicek-Moran, R. 2008. *Everyday science mysteries*. Arlington, VA: NSTA Press.

Metz, K. E. 1995. Reassessment of developmental constraints on children's science instruction. *Review of Educational Research* 65 (2): 93–127.

Yankee Publishing. *The old farmer's almanac,* published yearly since 1792. Dublin, NH: Yankee Publishing.

CHAPTER 1

THEORY BEHIND THE BOOK

We have all heard people refer to any activity that takes place in a science lesson as an "experiment." Actually, as science is taught today, true experiments are practically nonexistent. Experiments by definition test hypotheses, which are themselves virtually nonexistent in school science. A hypothesis, a necessary ingredient in any experiment, is a human creation developed by a person who has been immersed in a problem for a sufficient amount of time to feel the need to come up with a theory to explain events over which he or she has been puzzled.

However, it is quite common and proper for us to investigate our questions without proper hypotheses. Investigations can be carried out as "fair tests," which are possibly more appropriate for elementary classrooms, where children often lack the experience of creating a hypothesis in the true scientific mode. I recently asked a fourth grade girl what a "fair experiment" was and she replied, "It's an experiment where the answer is the one I expected." We cannot assume even at the fourth-grade level students are comfortable with controlling variables; it needs repeating.

A hypothesis is more than a guess. It will most often contain an "if… then…" statement, such as, "**If** I put a thermometer in a mitten and the temperature stays the same, **then,** perhaps the mitten did not produce heat." In school science, predictions should also be more than mere guesses or hunches, but rather based on experience and thoughtful consideration. Consistently asking children to give reasons for their predictions is a good

way to help them see the difference between guessing and predicting.

Two elements are often missing in most school science curricula: sufficient *time* to puzzle over problems that have some *real life applications*. It is much more likely that students will use a predetermined amount of time to "cover" an area of study—pond life, for example—with readings, demonstrations, and a field trip to a pond with an expert, topped off with individual or group reports on various pond animals and plants, complete with shoebox dioramas and giant posters. Or there may be a study of the solar system, with reports on facts about the planets and culminating with a class model of the solar system hung from the ceiling. These are naturally fun to do, but the issue is that there are seldom any real problems—nothing into which the students can sink their collective teeth into and use their minds to ponder, puzzle, hypothesize, and experiment.

You have certainly noticed that most science curricula have a series of "critical" activities in which students participate that supposedly lead to an understanding of a particular concept. In most cases, there is an assumption that students enter the study of a new unit with a common view or a common set of preconceptions about certain concepts and the activities will move the students closer to the accepted scientific view. This is a particularly dangerous assumption, since research shows that students enter into learning situations with a variety of preconceptions. These preconceptions are not only well ingrained in the students' minds but are exceptionally resistant to change. Going through the series of prescribed activities will have little meaning to students who have pre-conceptions that have little connection to the planned lessons, especially if the preconceptions are not recognized or addressed.

Bonny Shapiro, in her book, *What Children Bring to Light* (1994), points out in indisputable detail how a well-meaning science teacher ran his students through a series of activities on the nature of light without knowing that the students in the class all shared the misconception that seeing any object originates in the eye of the viewer and not from the reflection of light from an object into the eye. The activities were, for all intents and purposes, wasted, although the students had "solved the teacher" to the extent that they were able to fill in the worksheets and pass the test at the end of the unit—all the while doubting the critical concept that light reflecting from object to eye was the paramount fact and meaning of the act of seeing. *Solving the teacher* means that the students have learned a teacher's mannerisms, techniques, speech patterns, and teaching methods to the point that they can predict exactly what the teacher wants, what pleases or annoys her, and how to perform so the teacher believes her students have learned and understood the concepts she attempted to teach.

In her monograph *Inventing Density* (1986), Eleanor Duckworth says, "The critical experiments themselves cannot impose their own meanings. One has to have done a major part of the work already. One has to have developed a network of ideas in which to imbed the experiments." This may be the most important quote in this book!

How does a teacher make sure students develop a network of ideas in which to imbed the class activities? How does the teacher uncover student misconceptions about the topic to be studied? I believe that this book can offer some answers to these questions and offer some suggestions for remedying the problems mentioned above.

WHAT IS INQUIRY, ANYWAY?

There is probably no one definition of "teaching for inquiry," but at this time the acknowledged authorities on this topic have to be the National Research Council (NRC) and the American Association for the Advancement of Science (AAAS). After all, they are respectively the authors of the *National Science Education Standards* (1996) and the *Benchmarks for Science Literacy* (1993), upon which most states have based their curriculum standards. For this reason, I will use their definition, which I will follow throughout the book. The NRC, in *Inquiry and the National Science Education Standards: A Guide for Teaching and Learning* (2000), says that for real inquiry to take place in the classroom, the following five essentials must occur:

- Learner engages in scientifically oriented questions.
- Learner gives priority to evidence in responding to questions.
- Learner formulates explanations from evidence.
- Learner connects explanations to scientific knowledge.
- Learner communicates and justifies explanations. (p. 29)

In essence the NRC strives to encourage more learner self-direction and less direction from the teacher as time goes on during the school years. NRC also make it very clear that all science cannot be taught in this fashion. Science teaching that uses a variety of strategies is less apt to bore students and be more effective. Giving demonstrations, leading discussions, solving presented problems, and entering into a productive discourse about science are all viable alternatives. However, the NRC does suggest that certain common components should be shared by whichever instructional model is used:

- Students are involved with a scientific question, event, or phenomenon which connects with what they already know and creates a dissonance with their own ideas. In other words, they confront their preconceptions through an involvement with phenomena.
- Students have direct contact with materials, formulate hypotheses, test them and create explanations for what they have found.
- Students analyze and interpret data, and come up with models and explanations from these data.
- Students apply their new knowledge to new situations.
- Students engage in metacognition, thinking about their thinking, and review what they have learned and how they have learned it.

You will find opportunities to do all of the above by using these stories as motivators for your students to engage in inquiry-based science learning.

THE reasons FOR THIS BOOK

According to a summary of current thinking in science education in the journal *Science Education,* "one result seems to be consistently demonstrated: students leave science classes with more positive attitudes about science (and their concepts of themselves as science participants) when they learn science through inductive, hands-on techniques in classrooms where they're encouraged by a caring adult and allowed to process the information they have learned with their peers" (1993).

This book, and particularly the stories that lie within, provide an opportunity for students to take ownership of their learning and as stated in the quotation above, learn science in a way that will give them a more positive attitude about science and to process their learning with their classmates and teachers. Used as intended, the stories will require group discussions, hands-on and minds-on techniques, and a caring adult.

THE STORIES

These stories are similar to mystery tales but purposely lack the final chapter where the clever sleuth finally solves the mystery and tells the readers not only "whodunit," but how she knew. Because of the design of the tales in this book, the

students are challenged to become the sleuths and come up with likely "suspects" (the hypotheses or predictions) and carry out investigations (the experiments or investigations) to find out "whodunit" (the results). In other words, they write the final ending or perhaps endings. They are placed in a situation where they develop, from the beginning, "the network of ideas in which to imbed activities," as Duckworth suggests (1986, p. 39). The students are also the designers of the activities and therefore have invested themselves in finding the outcomes that make sense to them. I want them to have solved the problem rather than having solved the teacher. I do want to reemphasize however, that we should all be aware that successful students do spend energy in solving their teachers.

In one story ("Party Meltdown"), Kelsey and her friends wonder why ice left in a plastic bowl did not melt while ice left in a metal bowl did. They have to consider the differences in heat transfer between the two types of matter. Truly this is science as process and product. It also means that the students "own" the problem. This is what we mean by "hands-on, minds-on" science instruction. The teachers' belief in the ability of their students to own the questions and to carry out the experiments to reach conclusions, is paramount to the process. Each story has suggestions as to how the teachers can move from the story reading to the development of the problems, the development of the hypotheses and eventually the investigations that will help their students to come to conclusions.

Learning science through inquiry is a primary principle in education today. You might well ask, "instead of what?" Well, instead of learning science as a static or unchanging set of facts, ideas, and principles without any attention being paid to how these ideas and principles were developed. Obviously, we cannot expect our students to discover all of the current scientific models and concepts. We do however, expect them to appreciate the processes through which the principles are attained and verified. We also want them to see that science includes more than just what occurs in a classroom; that the everyday happenings of their lives are connected to science. Exploring the implications of friction, trying to repair a crooked garden swing, or wondering about how seeds can grow in a closed jar are only some of the examples of everyday life connected to science as a way of thinking and as a way of constructing new understandings about our world.

There are 15 stories in this book, each one focused on a particular conceptual area, such as thermodynamics, heat energy, melting or dissolving, the chemistry of cooking, astronomy, decomposition, and determining differences in reaction time. Each story can be photocopied and distributed to students to read and discuss or they can be read aloud to students and discussed by the entire class. During the discussion, it is ultimately the role of the teacher to help the students to find the problem or problems and then design ways to find out answers to the questions they have raised.

Most stories also include a few "distractors," also known as common misconceptions or alternative conceptions. The distractors are usually placed in the stories as opinions voiced by the characters who discuss the problematic situation. For example, in "Springtime in the Greenhouse: Planting Season," family members

argue over what is necessary for seeds to germinate. Each family member has his or her own preconception or misconception. The identification of these misconceptions is the product of years of research, and the literature documents the most common, often shared by both children and adults. Where do these common misconceptions come from and how do they arise?

DEVELOPMENT OF MENTAL MODELS

Until recently, educational practice has operated under the impression that children and adults come to any new learning situation without the benefit of prior ideas connected to the new situation. Research has shown that in almost every circumstance, learners have developed models in their mind to explain many of the everyday experiences they have encountered (Bransford, Brown, and Cocking 1999; Watson and Konicek 1990; Osborne and Fryberg 1985). Everyone has had experience with differences in temperature as they place their hands on various objects. Everyone has seen objects in motion and certainly has been in motion, either in a car, plane, or bicycle. Everyone has experienced forces in action, upon objects or upon themselves. Finally, each of us has been seduced into developing a satisfactory way to explain these experiences and to have developed a mental model, which explains these happenings to our personal satisfaction. Probably, most individuals have read books, watched programs on TV or in movie theaters, and used these presented images and ideas to embellish their personal models. It is even more likely that they have been in classrooms where these ideas have been discussed by a teacher or by other students. The film *A Private Universe* (Schnepps 1987), documents that almost all of the interviewed graduates and faculty of Harvard University showed some misunderstanding for either the reasons for the seasons, or for the reasons for the phases of the Moon. Many had taken high-level science courses either in high school or at the university.

According to the dominant and current learning theory called *constructivism*, all of life's experiences are integrated into the person's mind; they are accepted or rejected or even modified to fit existing models residing in that person's mind. Then, these models are used and tested for their usefulness in predicting outcomes experienced in the environment. If a model works, it is accepted as a plausible explanation; if not, it is modified until it does fit the situations one experiences. Regardless, these models are present in everyone's minds and brought to consciousness when new ideas are encountered. They may be in tune with current scientific thinking but more often they are "common sense science" and not clearly consistent with current scientific beliefs.

One of the reasons for this is that scientific ideas are often counterintuitive to everyday thinking. For example, when you place your hand on a piece of metal in a room, it feels cool to your touch. When you place your hand on a piece of wood in the same room it feels warmer to the touch. Many people will deduce that the temperature of the metal is cooler than that of the wood. Yet, if the objects have been in the same room for any length of time, their temperatures will be equal.

It turns out that when you place your hand on the metal, it conducts heat out of your hand quickly, thus giving the impression that it is cold. The wood does not conduct heat as rapidly as the metal and therefore 'feels' warmer than the metal. In other words, our senses have fooled us into thinking that instead of everything in the room being at room temperature, the metal is cooler than anything else. Therefore our erroneous conclusion is that metal objects are always cooler than other objects in a room. Indeed, if you go from room to room and touch many objects, your idea is reinforced and becomes more and more resistant to change.

These ideas are called by many names: *misconceptions, prior conceptions, children's thinking,* or *common sense ideas.* They all have two things in common. They are usually firmly embedded in the mind and they are highly resistant to change. Finally, if allowed to remain unchallenged these ideas will dominate a student's thinking, for example, about heat transfer, to the point that the scientific explanation will be rejected completely regardless of the method by which it is presented.

Our first impression is that these preconceptions are useless and must be quashed as quickly as possible. However, they are useful since they are the precursors of new thoughts and should be modified slowly toward the accepted scientific thinking. New ideas will replace old ideas only when the learner becomes dissatisfied with the old idea and realizes that a new idea works better than the old. It is our role to challenge these preconceptions and move learners to consider new ways of looking at their explanations and to seek ideas that work in broader contexts with more reliable results.

WHY STOries?

Why stories? Primarily, stories are a very effective way to get someone's attention. Stories have been used since the beginning of recorded history and probably long before that. Myths, epics, oral histories, ballads, dances, and such have enabled humankind to pass on the culture of one generation to the next, and the next, *ad infinitum.* Anyone who has witnessed story time in classrooms, libraries, or at bedtime knows the magic held in a well-written, well-told tale. They have beginnings, middles, and ends.

These stories begin like many familiar tales do: in homes or classrooms; with children interacting with siblings, classmates, or friends; with parents or other adults in family situations. But here the resemblance ends between our stories and traditional ones.

Science stories normally have a theme or a scientific topic that unfolds giving a myriad of facts, principles and perhaps a set of illustrations or photographs, which try to explain to a child the current understanding about the given topic. For years science books have been written as reviews of what science has constructed to the present. These books have their place in education, even though children often get the impression from these books that the information they have just read about appeared magically as scientists went about their work and "discovered"

truths and facts depicted in those pages. But as Martin and Miller (1990) put it: "The scientist seeks more than isolated facts from nature. The scientist seeks a *story* [emphasis mine]. Inevitably the story is characterized by a *mystery*. (Emphasis mine) Since the world does not yield its secrets easily, the scientist must be a careful and persistent observer."

As our tales unfold, discrepant events and unexpected results tickle the characters in the stories and stimulate their wonder centers making them ask, "What's going on here?" Most important of all, our stories have endings which are different than most. They are the mysteries that Martin and Miller talk about. They end with an invitation to explore and extend the story and to engage in inquiry.

These stories do not come with built-in experts who eventually solve the problem and expound on the solution. There is no Doctor Science who sets everybody straight in short order. Moms, dads, big sisters, brothers, and friends may offer opinionated suggestions ripe for consideration, or tests to be designed and carried out. It is the readers who are invited to become the scientists and solve the problem.

references

American Association for the Advancement of Science (AAAS).1993. *Benchmarks for science literacy.* New York: Oxford University Press.

Bransford, J. D., A. L. Brown, and R. R. Cocking, eds. 1999. *How people learn.* Washington, DC: National Academy Press.

Duckworth, E. 1986. *Inventing density.* Grand Forks, ND: Center for Teaching and Learning, University of North Dakota.

Martin, K., and E. Miller. 1990. Storytelling and science. In *Toward a whole language classroom: Articles from language arts, 1986–1989*, ed. B. Kiefer. Urbana, IL: National Council of Teachers of English.

National Research Council (NRC). 2000. *Inquiry and national science education standards: A guide for teaching and learning.* Washington, DC: National Academy Press.

Osborne, R., and P. Fryberg. 1985, *Learning in science: The implications of children's science.* Auckland, New Zealand: Heinemann.

Schneps, M. 1996. *The private universe project.* Harvard Smithsonian Center for Astrophysics.

Shapiro, B. 1994. *What children bring to light*, New York: Teachers College Press.

Watson, B., and R. Konicek. 1990, Teaching for conceptual change: Confronting children's experience. *Phi Delta Kappan* 71 (9): 680–684.

CHAPTER 2
USING THE BOOK AND THE STORIES

It is often difficult for overburdened teachers to develop lessons or activities that are compatible with the everyday life experiences of their students. A major premise of this book is that if students can see the real-life implications of science content, they will be motivated to carry out hands-on, minds-on science investigations and personally care about the results. Science educators have, for decades, emphasized the importance of science experiences for students that emphasize personal involvement in the learning process. I firmly believe that the use of open-ended stories that challenge students to engage in real experimentation about real science content can be a step toward this goal. Furthermore, I believe that students who see a purpose to their learning and experimentation are more likely to understand the concepts they are studying. I sincerely hope that the contents of this book will relieve the overburdened teacher from the exhausting work of designing inquiry lessons from scratch.

These stories feature children in natural situations at home, on the playground, at parties, in school, or in the outdoors. Students should identify with the story characters, to share their frustrations, concerns, and questions. The most important role for the adult is to help guide and facilitate investigations and to debrief activities with them and to think about their analyses of results and conclusions. The children often need help to go to the next level and to develop new questions and find ways of following these questions to a conclusion. Our philosophy of science education is based on the belief that children can and want to care enough about problems to make them their own. This should enhance and invigorate any curriculum. In short, students can begin to lead the curriculum and because of their personal interest in the questions that evolve from their activities, they will maintain interest for much longer than they would if they were following someone else's lead.

A teacher told me that one of her biggest problems is to get her students to "care" about the topics they are studying. She says they go through the motions but without affect. Perhaps this same problem is familiar to you. I hope that this book can help you to take a step toward solving that problem. It is difficult if not impossible to make each lesson personally relevant to every student. However, by focusing on everyday situations and highlighting kids looking at everyday phenomena, I believe that we can come closer to reaching student interests.

I strongly suggest the use of complementary books as you go about planning for inquiry teaching. Five special books are *Uncovering Student Ideas* (volumes 1, 2, 3, and 4) by Page Keeley et al., published by the NSTA Press and *Science Curriculum Topic Study* by Page Keeley, published by Corwin Press and NSTA. The multivolume *Uncovering Student Ideas* helps you to find out what kinds of preconceptions your students bring to your class. *Science Curriculum Topic Study* focuses on finding the background necessary to plan a successful standards-based unit. I would also strongly recommend that you find a copy of *Science Matters: Achieving Scientific Literacy*, by Robert Hazen and James Trefil. This book will become your reference for many scientific matters. It is written in a simple, direct and accurate manner and will give you the necessary background in the sciences when you need it. Finally please acquaint yourself with *Making Sense of Secondary Science: Research Into Children's Ideas* (Driver et al. 1994). The title of this book can be misleading to American teachers, because in Great Britain, anything above primary level is referred to as secondary. It is a compilation of the research done on children's thinking about science and is a must-have for teachers. Use it as a reference in looking for the preconceptions your students probably bring to your classroom.

In 1978, David Ausubel made one of the most simple but telling comments about teaching: "The most important single factor influencing learning is what the learner already knows; ascertain this, and teach him accordingly." The background material that accompanies each story is designed to help you to find out what your learners already know about your chosen topic and what to do with that knowledge as you plan. The above-mentioned books will supplement the materials in this book and deepen your understanding of teaching for inquiry.

How then, is this book set up to help you to plan and teach inquiry-based science lessons?

HOW THIS BOOK IS ORGANIZED

The stories are arranged in three sections. There are four for Earth systems science and technology, five for the biological sciences, and six for the physical sciences. There is a concept matrix at the beginning of each section that can be used to select a story most related to your content need. Following this matrix you will find the stories and the background material in separate chapters. Please note that the Earth systems science stories purposefully integrate the physical and biological sciences into science mysteries that focus on all aspects of everyday science related to the Earth sciences.

Each chapter, starting with Chapter 5, will have the same organizational format. First you will find the story, followed by background material for using the story. The background material will contain the following sections:

Purpose
This section describes the concepts and/or general topic that the story attempts to address. In short, it tells you where this story fits into the general scheme of science concepts. It may also place the concepts within a conceptual scheme of a larger idea.

Related Concepts

A concept is a word or combination of words that form a mental construct of an idea. Examples are *motion, reflection, rotation, heat transfer, acceleration*. Each story is designed to address a single concept but often the stories open the door to several concepts. You will find a list of possible related concepts in the teacher background material. You should also check the matrices of stories and related concepts.

Don't Be Surprised

In most cases, this section will include projections of what your students will most likely do and how they may respond to the story. The projections relate to the content but focus more on the development of their current understanding of the concept. The explanation will be related to the content but will focus more on the development of the understanding of the concept. There will be references made to the current alternative conceptions your students might be expected to bring to class. It may even challenge you to prepare for teaching by doing some of the projected activities yourself, so that you are prepared for what your students will bring to class.

Content Background

This material will be a very succinct "short course" on the conceptual material that the story targets. It will not, of course, be a complete coverage but should give you enough information to feel comfortable in using the story and planning and carrying out the lessons. In most instances, references to books, articles and internet connections will also help you in preparing yourself to teach the topic. It is important that you have a reasonable knowledge of the topic in order for you to lead the students through their inquiry. It is not necessary, however, for you to be an expert on the topic. Learning along with your students can help you to understand how their learning takes place and make you a member of the class team striving for understanding of natural phenomena.

Table 2.1.
Thematic Crossover Between Stories in This Book and *Uncovering Student Ideas in Science,*
Volumes 1–4

Story In this book	*Uncovering Student Ideas in Science*			
	Volume 1	Volume 2	Volume 3	Volume 4
The Coldest Time of the Day	n/a	n/a	What Is a Hypothesis?	Camping Trip

(continued to next page)

(continued from previous page)

Is the Earth Getting Heavier?	n/a	n/a	Rotting Apple; Earth's Mass; What Is a Hypothesis?	Is It a System?
The Moon Around the World	Gazing at the Moon; Going Through a Phase	Objects in the Sky; Emmy's Moon and Stars	What Is a Hypothesis?	Moonlight
Sunrise, Sunset	n/a	Darkness at Night; Objects in the Sky	Is It a Theory? Me and My Shadow; Where Do Stars Go? Summer Talk	Camping Trip
Lookin' at Lichens	n/a	Is It a Plant?	Does It Have a Life Cycle? Respiration	Is It "Fitter?"
Baking Bread	n/a	Chemical bonds	Doing Science; What Is a Hypothesis? Respiration	Is It Food? Is It a System?
Seeds Needs	Seedlings in a Jar	Is It a Plant/ Needs of Seeds; Is It Food for Plants?	Doing Science; What Is a Hypothesis? Does It Have a Life Cycle? Respiration	n/s
Reaction Time	Human Body Basics; Functions of Living Things	n/a	Doing Science; What Is a Hypothesis?	Is It a System?
Seedlings in a Jar	Functions of Living Things; Seedlings in a Jar	Is It Food for Plants?	Does It Have a Life Cycle?	Is It a system?
Sweet Talk	Is It Melting?	n/a	Doing Science; What Is a Hypothesis? Thermometer	Sugar Water

Cooling Off	The Mitten Problem; Objects and Temperature	Ice Cold lemonade; Mixing Water	Thermometer; What Is a Hypothesis?	Warming Water; Is It a System? Ice Water
Party Meltdown	Ice Cubes in a Bag	Ice Cold Lemonade; Freezing Ice	Is It a Theory? Thermometer	n/a
The Crooked Swing	n/a	n/a	Doing Science; What Is a Hypothesis?	Is It a System? Is It a Model?
Baking Cookies	n/a	n/a	Doing Science; What Is a Hypothesis?	Salt Crystals
Stuck	n/a	n/a	Apple on a Desk	Is It a System?

Related Ideas from the National Science Education Standards *(NRC) and* Benchmarks for Science Literacy *(AAAS)*

These two documents are considered to be the National Standards upon which most of the local and state standards documents are based. For this reason, the concepts listed for the stories are almost certainly the ones listed to be taught in your local curriculum. It is possible that some of the concepts are not mentioned specifically in the Standards but are clearly related. I suggest that you obtain a copy of *Curriculum Topic Study* (Keeley 2005), which will help you immensely with finding information about content, children's preconceptions, standards, and more resources. Even though it may not be mentioned specifically in each of the stories, you can assume that all of the stories will have connections to the Standards and Benchmarks in the area of Inquiry, Standard A.

Using the Story With Grades K–4 and 5–8

These stories have been tried with children of all ages. We have found that the concepts apply to all grade levels but at different levels of sophistication. Some of the characters in the stories have themes and characters that resonate better with one age group than another. However, the stories can be easily altered to appeal to an older or younger group by changing the characters to a more appropriate age or using slightly different age-appropriate dialog. The theme should be the same; just the characters and setting modified. Please read the suggestions for both grade levels.

As you may remember from the case study in the introduction, grade level is of little consequence in determining which stories are appropriate at which grade level. Both classes developed hypotheses and experiments appropriate to their developmental abilities. Second graders were satisfied to find out what happens to the length of a tree's shadow over a school year while the fifth grade class developed more sophisticated experiments involving length of day, direction of shadows over time and the daily length of shadows over an entire year. The main point here is that by necessity some stories are written with characters more appealing to certain age groups than others. Once again, I encourage you to read both the K–4 and 5–8 sections of Using the Story, because ideas presented for either grade level may be suited to your particular students.

There is no highly technical apparatus required. Readily available materials found in the kitchen, bathroom or garage will usually suffice. Each chapter includes with background information about the principles and concepts involved and a list of materials you might want to have available. These suggestions of ideas and materials are based upon our experience while testing these stories with children. While we know that classrooms, schools and children differ we feel that most childhood experiences and development result in similar reactions to explaining and developing questions about the tales. The problems beg for solutions and most importantly, create new questions to be explored by your young scientists.

Here you will find suggestions to help you to teach the lessons that will allow your students to become active inquirers develop their hypotheses and finally finish the story that you may remember was left open for just this purpose. I have not listed a step-by-step approach or set of lesson plans to accomplish this end. Obviously, you know your students, their abilities, their developmental levels, their learning abilities and disabilities better than anyone. You will find however, some suggestions and some techniques that we have found work well in teaching for inquiry. You may use them as written or modify them to fit your particular situation. The main point is that you try to involve your students as deeply as possible in trying to solve the mysteries posed by the stories.

Related NSTA Press Books and NSTA Journal Articles

Here, we will list specific books and articles from the constantly growing treasure trove of National Science Teacher Association (NSTA) resources for teachers. While our listings are not completely inclusive, you may access the entire scope of resources on the internet at *www.nsta.org/store*. Membership in NSTA will allow you to read all articles online.

References

References will be provided for the articles and research findings cited in the background section for each story.

Concept Matrices

At the beginning of each section you will find a concept matrix listing the concepts most related to each story. It can be used to select a story that matches your instructional needs.

FINAL WORDS

I was pleased find that Michael Padilla, Past President of NSTA, asked the same questions as I did when I decided to write a book that focused on inquiry. In the May 2006 edition of *NSTA Reports*, Mr. Padilla in his "President's Message" commented, "To be competitive in the future, students must be able to think creatively, solve problems, reason and learn new, complex ideas… [Inquiry] is the ability to think like a scientist, to identify critical questions to study; to carry out complicated procedures, to eliminate all possibilities except the one under study; to discuss, share and argue with colleagues; and to adjust what you know based on that social interaction." Further, he asks, "Who asks the question?…Who designs the procedures?…Who decides which data to collect?…Who formulates explanations based upon the data?…Who communicates and justifies the results?…What kind of classroom climate allows students to wrestle with the difficult questions posed during a good inquiry?"

I believe that this book speaks to these questions and that the techniques proposed here are one way to answer the above questions with, "the students do!" in the kind of science classroom this book envisions.

REFERENCES

Ausubel, D., J. Novak, and H. Hanensian. 1978. *Educational psychology: A cognitive view.* New York: Holt, Rinehart, and Winston.

Driver, R., A. Squires, P. Rushworth, and V. Wood-Robinson. 1994. *Making sense of secondary science: Research into children's ideas.* London and New York: Routledge Falmer.

Hazen, R., and J. Trefil. 1991. *Science matters: Achieving scientific literacy.* New York: Anchor Books.

Keeley, P. 2005. *Science curriculum topic study: Bridging the gap between standards and practice.* Thousand Oaks, CA: Corwin Press.

Keeley, P., F. Eberle, and C. Dorsey. 2008. *Uncovering student ideas in science, volume 3: Another 25 formative assessment probes.* Arlington, VA: NSTA Press.

Keeley, P., F. Eberle, and L. Farrin. 2005. *Uncovering student ideas in science, volume 1: 25 formative assessment probes.* Arlington, VA: NSTA Press.

Keeley, P., F. Eberle, and J. Tugel.2007. *Uncovering student ideas in science, volume 2: 25 more formative assessment probes.* Arlington, VA: NSTA Press.

Keeley, P., and J. Tugel. 2009. *Uncovering student ideas in science, volume 4: 25 new formative assessment probes.* Arlington, VA: NSTA Press.

Konicek-Moran, R. 2008. *Everyday science mysteries: Stories for inquiry-based science teaching.* Arlington, VA: NSTA Press.

Konicek-Moran, R. 2009. *More everyday science mysteries: Stories for inquiry-based science teaching.* Arlington, VA: NSTA Press.

Padilla, M. 2006. President's message. *NSTA Reports* 18 (9): 3.

CHAPTER 3
USING THIS BOOK IN DIFFERENT WAYS

Although the book was originally designed for use with K–8 students by teachers or adults in informal settings, it became obvious that a book containing stories and content material for teachers intent on teaching in an inquiry mode had other potential uses. I list a few of them below to show that the book has several uses beyond the typical elementary and middle school population in formal settings.

USING THE BOOK AS A CONTENT CURRICULUM GUIDE

When asked by the University of Massachusetts to teach a content course for a special master's degree program in teacher education, I decided to use *Everyday Science Mysteries* as one of several texts to teach content material. A major premise in the book is that students, when engaged in answering their own questions, will delve into a topic at a level commensurate with their intellectual development and learning skills. Therefore, even though the stories were designed for people younger than themselves, the students in the class were able to find questions to answer that were at a level of sophistication that challenged them.

During the fall 2007 semester this book was used as a text and curriculum guide for a class titled Exploring the Natural Sciences Through Inquiry at the University of Massachusetts in Amherst. The shortened version of the syllabus for the course follows:

Exploring the Natural Sciences Through Inquiry
EDUC 692 O
Fall 2007

Instructor: Dr. Richard D. Konicek, Professor Emeritus

Course Description:
This course is designed for elementary and middle school teachers who need, not only to deepen their content knowledge in the natural sciences, but also to understand how inquiry can be used in the elementary and middle school classroom. Natural sciences mean the Biological Sciences, Earth and Space Sciences and the Physical Sciences. Teachers will sample various topics from each of the above areas of science through inquiry techniques. The topics will be chosen from everyday phenomena such as Astronomy (Moon and Sun observations) Physics (motion, energy, thermodynamics, sound periodic motion, and Biology (botany, zoology, animal and plant behavior, evolution).

Course Objectives:
It is expected that each student will:
- Gain content background in each of the three areas of natural science.
- Be able to apply this content to their teaching methods.
- Develop questions concerning a particular phenomenon in nature.
- Design and carry out experiments to answer their questions.
- Analyze experimental data and draw conclusions.
- Consult various sources to verify the nature of their conclusions.
- Read scientific literature appropriate to their studies.
- Extend their knowledge to use with middle school children both in content and methodology.

Relationship to the Conceptual Framework of the School of Education:

Collaboration:	Teachers will work in collaborative teams during class meetings to acquire science content and pedagogical knowledge and skills.
Reflective Practice:	Teachers will develop and implement formative assessment probes with their students.
Multiple Ways of Knowing:	Teachers will share science questions and their methods of inquiry chosen to answer those questions.
Access, Equity, and Fairness:	Teachers reflect on student understandings based on students' stories.
Evidence-Based Practice:	Teachers will explore formative assessment through the use of probes.

Required Texts:
Hazen, R. M., and J. Trefil. 1991. *Science matters*. New York: Anchor Books.
Keeley, P., F. Eberle, and J. Tugel. 2007. *Uncovering student ideas in science: 25 more formative assessment probes, vol. 2*. Arlington, VA: NSTA Press. Konicek-

Moran, R. 2008. *Everyday science mysteries*. Arlington, VA: NSTA Press.

Resource Texts:

American Association for the Advancement of Science (AAAS). 2001. *Atlas of science literacy* (vol. 1). Washington, DC: Project 2061.

American Association for the Advancement of Science (AAAS). 2007. *Atlas of science literacy* (vol. 2). Washington, DC: Project 2061.

Driver, R., A. Squires, P. Rushworth, and V. Wood-Robinson. 1994. *Making sense of secondary science*. London: Routledge-Falmer.

Keeley, P., F. Eberle, and L. Farrin. 2005. *Uncovering student ideas in science, vol. 1*. Arlington, VA: NSTA Press.

Topics To Be Investigated in Volume One:

Everyday Science Mysteries is organized around stories. The core concepts related to the National Science Education Standards developed by the National Research Council in 1996 are the basis for the concept selection. The story titles and related core concepts are shown in the matrices below.

Earth Systems Science

Core Concepts	Stories				
	Moon Tricks	Where Are the Acorns?	Master Gardener	Frosty Morning	The Little Tent That Cried
States of Matter			X	X	X
Change of State			X	X	X
Physical Change			X	X	X
Melting			X	X	
Systems	X	X	X	X	X
Light	X	X			
Reflection	X	X		X	
Heat Energy			X	X	X
Temperature				X	X
Energy			X	X	X
Water Cycle				X	X
Rock Cycle			X		
Evaporation				X	X
Condensation				X	X
Weathering			X		
Erosion			X		
Deposition			X		
Rotation/Revolution	X	X			
Moon Phases	X				
Time	X	X			

Physical Sciences

Core Concepts	Stories				
	Magic Balloon	Bocce Anyone?	Grandfather's Clock	Neighborhood Telephone Service	How Cold Is Cold?
Energy	X	X	X	X	X
Energy Transfer	X	X	X	X	X
Conservation of Energy		X			X
Forces	X	X	X		
Gravity	X	X	X		
Heat	X				X
Kinetic Energy		X	X		
Potential Energy		X	X		
Position and Motion		X	X		
Sound				X	
Periodic Motion			X	X	
Waves				X	
Temperature	X				X
Gas Laws	X				
Buoyancy	X				
Friction		X	X		
Experimental Design	X	X	X	X	X
Work		X	X		
Change of State					X
Time		X	X		

Biological Sciences

Core Concepts	Stories				
	About Me	Bugs	Dried Apples	Seed Bargains	Trees From Helicopters
Animals	X	X			
Classification		X	X	X	X
Life Processes	X	X	X	X	X
Living Things	X	X	X	X	X
Structure and Function		X	X		X
Plants			X	X	X
Adaptation		X			X

NATIONAL SCIENCE TEACHERS ASSOCIATION

Genetics/ Inheritance	X		X	X	X
Variation	X		X	X	X
Evaporation			X		
Energy		X	X	X	X
Systems	X	X	X		X
Cycles	X	X	X	X	X
Reproduction	X	X	X	X	X
Inheritance	X	X	X		X
Change		X	X		
Genes	X		X		X
Metamorphosis		X			
Life Cycles		X	X		X
Continuity of Life	X	X	X	X	X

Assignments:

Astronomy (25%): Everyone will be expected to explore the daytime astronomy sequence, which will aim to develop models of the Earth, Moon, and Sun relationships. Students will keep a Moon journal and Sun shadow journal over the course of the semester, which they will turn in periodically.

Topics (50%): In addition, students will pick at least two topics from each of the Earth, Physical and Biological areas for study during the semester. Students will come up with a topic question and do an investigation or experiment regarding the topic questions posed. (For example: Are there acorns that do not need a dormancy period before germinating?) These questions and experiments will be shared with the class as they progress so that all students will either be directly involved in learning about the content or indirectly involved by listening to reports and critiquing those reports. In addition to the experiments, students will (1) involve their students in their experiments/investigations and (2) design and give formative assessment probes to their students to find out what knowledge they already possess. Students will be graded on their experimental designs, their presentations of their data and upon their conclusions. I will develop a rubric with the students that will address the goals stated above and their values to be calculated for their grades.

Attendance/Participation (25%): Attendance at all course meetings is required.

References for Course Development:

American Association for the Advancement of Science (AAAS).1993. *Benchmarks for science literacy.* New York: Oxford University Press.

Ausubel, D., J. Novak, and H. Hanensian. 1978. *Educational psychology: A cognitive view.* New York: Holt, Rinehart and Winston.

Bransford, J. D., A. L. Brown, and R. R. Cocking, eds. 1999. *How people learn.* Washington, DC: National Academy Press.

Duckworth, E. 1986. *Inventing density.* Grand Forks, ND: Center for Teaching and Learning, University of North Dakota.

Driver, R., A. Squires, P. Rushworth, and V. Wood-Robinson. 1994. *Making sense of secondary science: Research into children's ideas.* London and New York: Routledge Falmer.

Hazen, R., and J. Trefil. 1991. *Science matters: Achieving scientific literacy.* New York: Anchor Books.

Keeley, P. 2005. *Science curriculum topic study: Bridging the gap between standards and practice.* Thousand Oaks, CA: Corwin Press.

Keeley, P., F. Eberle, and L. Farrin. 2005. *Uncovering student ideas in science: 25 formative assessment probes* (vol. 1). Arlington, VA: NSTA Press.

Keeley, P., F. Eberle, and J. Tugel. 2007. *Uncovering student ideas in science: 25 more formative assessment probes* (vol. 2). Arlington, VA: NSTA Press.

Konicek-Moran, R. 2008. *Everyday science mysteries.* Arlington, VA: NSTA Press.

Martin, K., and E. Miller. 1990. Storytelling and science. In *Toward a whole language classroom: Articles from language arts,* ed. B. Kiefer, 1986–1989. Urbana, IL: National Council of Teachers of English.

National Research Council (NRC). 2000. *Inquiry and national science education standards: A guide for teaching and learning.* Washington, DC: National Academy Press.

Osborne, R., and P. Fryberg. 1985. *Learning in science: The implications of children's science.* Auckland, New Zealand: Heinemann.

Scnepps, M. 1996. *The private universe project.* Washington, DC: Harvard Smithsonian Center for Astrophysics.

Shapiro, B. 1994. *What children bring to light.* New York: Teachers College Press.

Watson, B., and R. Konicek. 1990. Teaching for conceptual change: Confronting children's experience. *Phi Delta Kappan* May: 680–684.

The course was taught as a graduate course for teachers or prospective teachers of elementary or middle students. The course could be classified as a content/pedagogy class for teachers who had minimal science backgrounds as well as minimal skills in teaching for inquiry. My premise was that if teachers would learn content through inquiry techniques, they would be convinced of their efficacy as a learning techniques and would be likely to use them to teach content, in their own classes. As it turned out, those teacher-students who had classes of their own and were full time teachers did work on their projects with their students with very satisfactory results according to the teachers. As a result, both teachers and students were learning science content through inquiry techniques. Because of the fact that the teachers in the class were completing an assignment, they were able to be honest with their students about not knowing the outcome of their investigations. This is

often a problem with teachers who are afraid to admit that they are learning along with the students. In this case, the students were excited about learning along with their teachers and vice versa. Teachers with classrooms were also able to develop rubrics with their students for the grading of their explorations and therefore were involved with some metacognition as well.

As a result of this small foray into the use of the book in this manner, I am convinced that the book can be used as a content guide for undergraduate and graduate content oriented courses for teachers. As noted in the syllabus, the use of other supplementary texts for content and pedagogy add to the strength of the course in preparing teachers to use inquiry techniques and to learn content themselves. With the use of the Internet, very little information is hidden from anyone with minimum computer skills. Unlike many survey courses chosen by teachers who are science-phobic, this course did not attempt to cover a great number of topics but to teach a few topics for understanding. The basic premise of this author is that when deciding between coverage and understanding science topics and concepts, understanding wins every time. It is well known that our current curriculum in the United States has been faulted for being a mile wide and an inch deep. High stakes testing seems to also add to the problem since almost all teachers whom I have interviewed over the last few years are reluctant to teach for understanding using inquiry methods because teaching for understanding takes more time and does not allow for coverage of the almost infinite amount of material which might appear on a standardized tests. Thus, student misconceptions are seldom addressed and continue to persist even though students can do reasonably well on teacher-made tests and assessment tools and still hold on to their misconceptions. See Bonnie Shapiro's book, *What Children Bring to Light*.

USING THIS BOOK AS A RESOURCE BOOK FOR SCIENCE METHODS COURSES IN TEACHING PREPARATION PROGRAMS

Traditionally, science methods courses in the United States are taught to classes mainly composed of science-phobic students. One of the main goals of science methods courses is to make students comfortable with science teaching and to help students develop skills in teaching science to youngsters using a hands-on, minds-on approach. Unfortunately, a great many students come to these methods courses with a minimum of science content courses and many of those are either survey (nonlaboratory) courses or courses taught in large lecture format. In 12–13 weeks, methods instructors are expected to convert these students into confident, motivated teachers who are familiar with techniques that promote inquiry learning among their students. Having taught this type of course to undergraduates and career-changing graduate students for over 30 years, I have found that making students comfortable with science is the first goal. This is often accomplished by assigning students science tasks that can be accomplished with a minimum of stress and with a maximum of success. Second, I try to instill the ideas commensurate with the nature of science as a discipline. Third, I find that it is often necessary to teach a little content for those who are rusty and need to clarify some of their own misconceptions. Lastly, but not least important, I try to acquaint them with resources in the field so that they know what is available to them as they enter their teaching careers. Obviously, here is an opportunity to acquaint them with current information about the learners themselves, how they learn and how to teach for inquiry.

As a final assignment for my methods classes, I assign the students the task of writing an everyday science mystery and a paper to accompany it, which will describe how they will use the story to teach a concept using the inquiry approach. The results have far exceeded what I had been receiving from the typical lesson plan used by others and me through the years. This book would not only provide the text on teaching science found in the early chapters but would provide a model for producing everyday science mysteries for topics of the students' choices.

Use For Homeschool Programs

Homeschooling parents have a great many resources at their disposal, as any internet search will show. Curricular suggestions and materials are available for those parents and children who choose to conduct their education at home. Science is one of those subjects that might be difficult for many parents whose science backgrounds are a bit weak or outdated. Parents and children working together to solve a story driven mystery could use this book easily. The connections to the National Standards and the Bookmarks in science also help in making sure that the homeschooling curriculum is uncovering the nationally approved scientific concepts. Parents would use the book just as any teacher would use it except there would be fewer opportunities for class discussions and the parents would have to do a bit more discussion with their children to solidify their understanding of their investigations.

reference

Shapiro, B. 1994. *What children bring to light.* New York: Teachers College Press.

CHAPTER 4
SCIENCE AND LITERACY

While heading into the final chapter before launching into the stories, I couldn't resist introducing you to a piece of literature that is seldom read except by English majors. The quotation that follows is from Irish novelist James Joyce in his classic book *Ulysses*, written in 1922:

> Where was the chap I saw in that picture somewhere? Ah, in the dead sea, floating on his back, reading a book with a parasol open. Couldn't sink if you tried: so thick with salt. Because the weight of the water, no, the weight of the body in the water is equal to the weight of the. Or is it the volume is equal to the weight? It's a law something like that. Vance in High school cracking his fingerjoints, teaching. The college curriculum. Cracking curriculum. What is weight really when you say weight? Thirtytwo feet per second, per second. Law of falling bodies: per second, per second. They all fall to the ground. The earth. It's the force of gravity of the earth is the weight. (p.73)

In the novel, Joyce's main character Bloom recalls a picture of someone floating in the Dead Sea, and tries to recall the science behind it. Have you or have you observed others who, while trying to recall something scientific, resorted to a mishmash of scientic knowledge, half-remembered and garbled? (For this foray into literature, I am indebted to Michael J. Reiss who called my attention to this passage in an article of his in *School Science Review*).

In his school days, Bloom seems to have been fascinated both with the curriculum and the teacher in his physics class. However, Bloom's memory of the science behind buoyancy runs the gamut from unrelated science language pouring out of his memory bank to visions of his teacher cracking his finger joints. Unfortunately, even today, this might well be the norm rather than the exception. This phenomenon is exactly what we are trying to avoid in our modern pedagogy and now leads us to the main point of this chapter.

There are many ways of connecting literacy and science. We shall look briefly at the research literature and find some ideas that will make the combination of literacy and science not only worthwhile but also essential for learning.

LITERACY AND SCIENCE

In pedagogical terms there are differences between scientific literacy and the curricular combination of science and literacy, but perhaps they have more in common than one might expect. *Scientific literacy* is the ability to understand scientific concepts so that they have a personal meaning in everyday life. In other words, a scientifically literate population can use their knowledge of scientific principles in situations other than those in which they learned them. For example, I would consider people scientifically literate if they were able to use their understanding of ecosystems and ecology to make informed decisions about saving wetlands in their community. This is of course, what we would hope for in every aspect of our educational goals regardless of the subject matter. *Literacy* refers to the ability to read, write, speak, and make sense of text. Since most schools emphasize reading, writing and mathematics, they often take priority over all other subjects in the school curriculum. How often have I heard teachers say that their major responsibility is reading and math, and that there is no time for science? But there is no need for competition for the school day. I believe that this misconception is caused by the lack of understanding of the synergy created by integration of subjects. In *synergy*, you get a combination of skills that surpasses the sum of the individual parts.

So what does all of this have to do with teaching science as inquiry? There is currently a strong effort to combine science and literacy. One reason is that there is a growing body of research that stresses the importance of language in learning science. "Hands-on" science is nothing without its "minds-on" counterpart. I am fond of reminding audiences that a food fight is a hands-on activity, but one does not learn much through mere participation, except perhaps the finer points of the aerodynamic properties of Jell-O. The understanding of scientific principles is not imbedded in the materials themselves or in the manipulation of these materials. Discussion, argumentation, discourse of all kinds, group consensus and social interaction—all forms of communication are necessary for students to make meaning out of the activities in which they have engaged. And these require *language* in the form of writing, reading, and particularly speaking. They require that students think about their thinking—that they hear their own and others' thoughts and ideas spoken out loud and perhaps eventually see them in writing to make sense of what they have been doing and the results they have been getting in their activities. This is the often forgotten "minds-on" part of the "hands-on, minds-on" couplet. Consider the following:

> In schools, talk is sometimes valued and sometimes avoided, but—and this is surprising—talk is rarely taught. It is rare to hear teachers discuss their efforts to teach students to talk well. Yet talk, like reading and writing, is a major motor—I could even say the major motor—of intellectual development. (Calkins 2000, p.226)

For a detailed and very useful discussion of talk in the science classroom, I refer you to Jeffrey Winokur and Karen Worth's chapter, "Talk in the Science Classroom: Looking at What Students and Teachers Need to Know and Be Able to Do" in *Linking Science and Literacy in the K–8 Classroom* (2006). Also check out Chapter 8 in this book. There is also recent evidence that ELL learners gain a great deal from talking, in both their science learning and new language acquisition (Rosebery and Warren 2008).

Linking inquiry-based science and literacy has strong research support. First, the conceptual and theoretical work of Padilla and his colleagues suggest that

inquiry science and reading share a set of intellectual processes (e.g., observing, classifying, inferring, predicting, and communicating) and that these processes are used whether the student is conducting scientific experiments or reading text (Padilla, Muth, and Padilla 1991). Helping children become aware of their thinking as they read and investigate with materials will help them understand and practice more *metacognition*.

You, the teacher, may have to model this for them by thinking out loud yourself as you view a phenomenon. Help them to understand why you spoke as you did and why it is important to think about your process of thinking. You may say something like, "I think that warm weather affects how fast seeds germinate. I think that I should design an experiment to see if I am right." Then later, "Did you notice how I made a prediction that I could test in an experiment?" Modeling your thinking can help your students see how and why the talk of science is used in certain situations.

Science is about words and their meanings. Postman made a very interesting statement about words and science. He said "Biology is not plants and animals. It is language about plants and animals.... Astronomy is not planets and stars. It is a way of talking about planets and stars" (1979, p.165). To emphasize this point even further, I might add that science is a language, a language that specializes in talking about the world and being in that world we call science. It has a special vocabulary and organization. Scientists use this vocabulary and organization when they talk about their work. Often, it is called "discourse" (Gee 2004). Children need to learn this discourse when they present their evidence, when they argue the fine points of their work, evaluate their own and others' work and refine their ideas for further study.

Students do not come to you with this language in full bloom; in fact the seeds may not even have germinated. They attain it by doing science and being helped by knowledgeable adults who teach them about controlling variables, conducting fair tests, having evidence to back up their statements, and using the processes of science in their attempts at what has been called "first hand inquiry" (Palincsar and Magnusson 2001). This is inquiry that uses direct involvement with materials, or in other more familiar words, the hands-on part of scientific investigation. The term *second hand investigations* refers to the use of textual matter, lectures, reading data, charts, graphs, or other types of instruction that do not feature direct contact with materials. Cervetti et al. (2006) put it so well:

> [W]e view firsthand investigations as the glue that binds together all of the linguistic activity around inquiry. The mantra we have developed for ourselves in helping students acquire conceptual knowledge and the discourse in which that knowledge is expressed (including particular vocabulary) is "read it, write it, talk it, do it!"—and in no particular order, or better yet, in every possible order. (p. 238)

So you can see that it is also important that the students talk about their work; write about their work; read about what others have to say about the work they are doing, in books or via visual media; and take all possible opportunities to document their work in a way that is useful to them in looking back at what they have found out about their work.

THE LanGuaGe OF science

Of course, writing, talking, and reading in the discipline of science is different than other disciplines. For example, science writing is simple and focuses on the evidence obtained to form a conclusion. But science includes things other than just verbal language. It includes tactile, graphic, and visual means of designing studies, carrying them out, and communicating the results to others. Also important is that science has many unfamiliar words; many common words such as *work, force, plant food, compound,* and *density* have different meanings in the real world of the student but have precise and often counterintuitive meanings in science. For example, if you push against a car for 30 minutes until perspiration runs off your face, you feel as though you have "worked" hard even though the car has not budged a centimeter. In physics, unless the car has moved, you have done no work at all. We tell students that plants make their own food and then show them a bag of "plant food." We tell children to "put on warm clothes," yet the clothes have nothing to do with producing warmth.

Students have to change their way of communicating when they study science. They must learn new terminology and clarify old terms in scientific ways. We as teachers can help in this process by realizing that we are not just science teachers but also language teachers. When we talk of scientific things we talk about them in the way the discipline works. We should not avoid scientific terminology but try to connect it whenever possible to common metaphors and language. We should use pictures and stories.

We need also to know that science contains many words that ask for thought and action on the part or the students. Sentences with words like *compare, evaluate, infer, observe, modify,* and *hypothesize* prompt students to solve problems. We can only teach good science by realizing that language and intellectual development go hand in hand and that one without the other is mostly meaningless.

SCIENCE NOTEBOOKS

Many science educators have lately touted science notebooks as an aid to students involving themselves more in the discourse of science (Klentschy 2005; Campbell and Fulton 2003). Their use has also shown promise in helping English language learners (ELLs) in the development of language skills as well as learning science concepts and the nature of science (Klentschy 2005).

Science notebooks differ from science journals and science logs in that they are not merely for recording data (logs) or reflections of learning (journals), but

are meant to be used continuously for recording experimentation, designs, plans, thinking, vocabulary, and concerns or puzzlement. The science notebook is the recording of past, and present thoughts and predictions and are unique to each student. The teacher makes sure that the students have ample time to record events and to also ask for specific responses to such questions as, "What still puzzles you about this activity?"

For specific ideas for using science notebooks and for information on the value of using the notebooks in science, see *Science Notebooks: Writing About Inquiry* by Brian Campbell and Lori Fulton (2003) and "Science Notebook Essentials: A Guide to Effective Notebook Components," an article in *Science and Children*, by Michael Klentschy (2005).

You can assume that science notebooks are a given in what I envision as an inquiry-oriented classroom. While working in an elementary school years ago, I witnessed some minor miracles of children writing to learn. The most vital lesson for us as teachers was the importance of asking children to write each day about something that still confused them. The results were remarkable. As we read their notebooks, we witnessed their metacognition, and their solutions through their thinking "out loud" in their writing.

The use of science notebooks should be an opportunity for the students to record their mental journey through their activity. Using the stories in this book, the science notebook would include the specific question that the student is concerned with, the lists of ideas and statements generated by the class after the story is read, pictures or graphs of data collected by the student and class, and perhaps the final conclusions reached by the student or class as they try to solve the mystery presented by the story.

Let us imagine that your class has reached a conclusion to the story they have been using and have reached consensus on that conclusion. What options are open to you as a teacher for asking the students to finalize their work? At this juncture, it may be acceptable to have the students actually write the "ending" to the story or write up the conclusions in a standard lab report format. The former method, of course, is another way of actually connecting literacy and science. Many teachers prefer to have their students at least learn to write the "boiler plate" lab reports, just to be familiar with that method, while others are comfortable with having their students write more anecdotal kinds of reports. My experience is that when students write their conclusions in an anecdotal form, while referring to their data to support their conclusions, I am more assured that they have really understood the concepts they have been chasing rather than filling in the blanks in a form. In the end, it is up to you, the classroom teacher, to decide. Of course, it could be done both ways.

As mentioned earlier, a major factor in designing these stories and follow-up activities is based upon one of the major tenets of a philosophy called *constructivism*. That tenet is that knowledge is constructed by individuals in order to make sense out of the world in which they live. If we believe this, then the knowledge that each individual brings to any situation or problem must be factored into the way that person tries to solve that problem. By the same token, it is most important to realize that the *identification* of the problem and the way the problem is *viewed* are also factors determined by each individual. Therefore it is vital that the adult facilitator encourage the students to bring into the open, orally and in writing, those ideas they already have about the situation being discussed. In bringing these preconceptions out of hiding, so to speak, all of the children and the teacher can begin playing with all of the cards exposed and alternative ideas about topics can be addressed. Data can be then analyzed openly without any hidden agendas in childrens' minds to sabotage learning. You can find more about this process in the series *Uncovering Student Ideas in Science: 25 Formative Assessment Probes*, vols.1–4 (2005, 2007, 2008, 2009).

The stories also point out that science is a social, cultural, and therefore human enterprise. The characters in our stories usually enlist others in their investigations, their discussions and their questions. These people have opinions and hypotheses and are consulted, involved or drawn into an active dialectic. Group work is encouraged which in a classroom would suggest cooperative learning. At home, siblings and parents may become involved in the activities and engage in the dialectic as a family group.

The stories can also be read to the children. In this way children can gain more from the literature than if they had to read the stories by themselves. A child's listening vocabulary is usually greater than his or her reading vocabulary. Words that are somewhat unfamiliar to them can be deduced by the context in which they are found. Or, new vocabulary words can be explained as the story is read. We have found that children are always ready to discuss the stories as they are read and therefore become more involved as they take part in the reading. So much the better because getting involved is what this book is all about; getting involved in situations that beg for problem finding, problem solving, and construction of new ideas about science in everyday life.

HELPING YOUR STUDENTS DURING INQUIRY

How much help should you give to your students as they work through the problem? A good rule of thumb is that you can help them as much as you think necessary as long as the children are still finding the situation problematic. In other words, the children should not be following your lead but their own lead. If some of these leads end up in dead ends, then that aspect of scientific investigation is part of their experience too. Science is full of experiences which are not productive. If children read popular accounts of scientific discovery, they could get the impression that the scientist gets up in the morning, says, "What will I discover today?" and then sets off on a clear, straight path to an elegant conclusion before suppertime rolls around. Nothing could be further from the truth! But it is very important to note that a steady diet of frustration can dampen students' enthusiasm for science.

Dead ends can be viewed as signaling a need to develop a new plan or ask the question in a different way. Most important, dead ends should not be looked upon

as failures. They are more like opportunities to try again in a different way with a clean slate. The adult's role is to keep a balance so that motivation is maintained and interest continues to flourish. Sometimes this is more easily accomplished when kids work in groups. Most often nowadays, scientists work in teams and use each other's expertise in a group process,

Many people do not understand that the scientific process includes luck, personal idiosyncrasies, and feelings, as well as the so-called "scientific method." The term "scientific method" itself sounds like a recipe guaranteed to produce success. The most important aid you can provide for your students is to help them maintain their confidence in their ability to do problem solving using all of their ways of knowing. They can use metaphors, visualizations, drawings, or any other method with which they are comfortable to develop new insights into the problem. Then they can set up their study in a way that reflects the scientific paradigm including a simple question, controlling variables and isolating the one variable they are testing.

Next, you can help them to keep their experimental designs simple and carefully controlled. Third, you can help them to learn keep good data records in their science notebooks. Most students don't readily see the need for this last point, even after they have been told. They don't see the need because the neophyte experimenter has not had much experience with collecting usable data. Until they realize that unreadable data or necessary data not recorded can cause a problem, they see little use for them. The problem is that they don't see it as a problem. Children don't see the need for keeping good shadow length records because they are not always sure what they are going to do with them in a week or a month from now. If they are helped to see the reasons for collecting data and that these data are going to be evidence of a change over time, then they will see the purpose of being able to go back and revisit the past in order to compare it to the present. In this way they can also see the reasons for keeping a log in the first place.

In experiences we have had with children, forcing them to use prescribed data collection worksheets has not helped them to understand the reasons for data collection at all and in some cases has actually caused more confusion or amounted to little more than busy work. On one occasion while circulating around a classroom where children were engaged in a worksheet-directed activity, an observer asked a student what she was doing. The student replied without hesitation, "step three." Our goal is to empower students engaged in inquiry to the point where they are involved in the activity at a level where all of the steps, including step three, are designed by the students themselves and for good reason--to answer their own questions in a logical, sequential, meaningful manner. We believe it can be done but it requires patience on the part of the adult facilitators and faith that the children have the skills to carry out such mental gymnastics, with a little help from their friends and mentors.

One last word about data collection. After spending years being a scientist and working with scientists, one common element stands out for me. Scientists keep on their person a notebook that is used numerous times during the day to record interesting items. The researcher may come across some interesting data that may not seem directly connected to the study at the time but he or she makes some notes about it anyway because that entry may come in handy in the future. Memory is viewed as an ephemeral thing, not to be trusted. Scientists' notebooks are a treasured and essential part of the scientific enterprise. In some cases they have been considered legal documents and used as such in courts of law. There is an ethical expectation that scientists record their data honestly. Many times, working with my mentor, biologist Skip Snow in the Everglades National Park Python Project, I have seen Skip refer to previous entries when confronted with data that he thinks may provide a clue to a new line of investigation. Researchers don't leave home without notebooks.

WORKING WITH ENGLISH LANGUAGE LEARNER (ELL) POPULATIONS

Now, suppose that members of your class are from other cultures and have a limited knowledge of the English language. Of what use is inquiry science with such a population and how can you use the discipline to increase both their language learning and their science skills and knowledge?

First of all, let's take a look at the problems associated with learning with the handicap of limited language understanding. Lee (2005) in her summary of research on ELL students and science learning, points to the fact that students who are not from the dominant culture are not aware of the rules and norms of that culture. Some may come from cultures in which questioning (especially of elders) is not encouraged and where inquiry is not supported. Obviously, to help these children cross over from the culture of home to the culture of school, the rules and norms of the new culture must be explained carefully and visibly, and the students must be helped to take responsibility for their own learning. You can find specific help in a recent NSTA publication by Ann Fathman and David Crowther (2006) entitled *Science for English Language Learners: K–12 Classroom Strategies*. Also very helpful is another NSTA publication, *Linking Science and Literacy in the K–8 Classroom*. Chapter 12, "English Language Development and the Science-Literacy Connection"(Douglas, Klentschy, and Worth 2006). Add to this array of written help two more books: *Teaching Science to English Language Learners: Building on Students' Strengths* (Roseberry and Warren 2008) and *Science for English Language Learners: K–12 Classroom Strategies* (Fathman and Crowther 2006). Finally, an article from *Science and Children* (Buck 2000) entitled "Teaching Science to English-as-Second Language Learners" has many useful suggestions for working with ELL students.

I can summarize as best as I can a few ideas and will also put them into the teacher background sections when appropriate.

Experts agree that vocabulary building is very important for ELL students. You can focus on helping these students identify objects they will be working with

in their native language and in English. These words can be entered in science notebooks. Some teachers have been successful in using a teaching device called a "working word wall." This is an ongoing poster with graphics and words that are added to the poster as the unit progresses. When possible, real items or pictures are taped to the poster. This is visible for constant review and kept in a prominent location, since it is helpful for all students, not just the ELL students.

Many teachers suggest that the group work afforded by inquiry teaching helps ELL students understand the process and the content. Pairing ELL students with English speakers will facilitate learning since often students are more comfortable receiving help from peers than from the teacher. They are more likely to ask questions of peers as well. It is also likely that explanations from fellow students may be more helpful, since they'll probably explain things in language more suitable to those of their own age and development.

Use the chalkboard or whiteboard more often. Connect visuals with vocabulary words. Remember that science depends upon the language of discourse. You might also consider inviting parents into the classroom so that they can witness what you are doing to help their children to learn English and science. Spend more time focusing on the process of inquiry so that the ELL students will begin to understand how they can take control over their own learning and problem solving.

The SIOP model (Echevarria, Vogt, and Short 2004) has been earning popularity lately with teachers who are finding success in teaching science to ELL students. SIOP is an acronym for Sheltered Instruction Observation Protocol. It emphasizes hands-on/minds-on types of science activities that require ELL students to interact with their peers using academic English. You can reach the SIOP Institute website at *www.siopinstitute.net*. While it is difficult to summarize the model succinctly, the focus is on melding the use of academic language with inquiry-based instruction. Every opportunity to combine activity and inquiry should be taken and all of the many types of using language be stressed. This would include writing, speaking, listening, and reading. There is also a strong emphasis on ELL students being paired with competent English language speakers so that they can listen and practice using the vocabulary with those students who have a better command of the language.

In short, the difference between most other ESL programs and Sheltered Instruction is that in the latter, the emphasis is on connecting the content area learning and language learning in such a way that they enhance each other rather than focusing on either the content or the language learning as separate entities. In many programs it is assumed that ELL students cannot master the content of the various subjects because of their lack of language proficiency. Sheltered Instruction assumes that given more opportunities to speak, write, read, talk and listen in the context of any subject's language base, ELL students can master the content as well as the academic language that goes with the content.

Teachers also need to be more linguistically present during classroom management tasks. They need to talk with students to make sure they are interpreting their inquiry tasks and learning how to explain their observations and conclusions in their new language. The teacher's role includes making sure students are focused by reminding them to write things down and to help them discuss their findings in English. As I said before, it is not only the ELL students who need to work on their academic language but all students who need to learn that science has a way of using language and syntax that is different than other disciplines. All students can benefit from being considered Science Language Learners.

And now, on to the stories which I hope will inspire your students to become active inquirers and enjoy science as an everyday activity in their lives.

references

Buck, G. A. 2000. Teaching science to English-as-second language learners. *Science and Children* 38 (3): 38–41.

Calkins, L. M. 2000. *The art of teching reading.* Boston: Allyn and Bacon.

Campbell, B., and L. Fulton. 2003. *Science notebooks: Writing about inquiry.* Portsmouth,NH:Heinemann.

Cervetti,G. N., P. D. Pearson, M. Bravo, and J. Barber. 2006. Reading and writing in the service of inquiry-based science. In *Linking Science and Literacy in the K–8 classroom,* ed. R. Douglas, M. Klentschy, and K. Worth, 221–244. Arlington, VA: NSTA Press.

Douglas, R., M. P. Klentschy, and K. Worth, eds. 2006. *Linking science and literacy in the K–8 classroom.* Arlington, VA: NSTA Press.

Echevarria, J., M. E. Vogt, and D. Short. 2000. *Making content compresible for English language learners: The SIOP model.* Needham Heights. MA: Allyn and Bacon.

Fathman, A., and D. Crowther. 2006. *Science for English language learners: K–12 classroom strategies.* Arlington, VA: NSTA Press.

Gee, J. P. 2004. Language in the science classroom: Academic social languages as the heart of school-based literacy. In *Crossing borders in literacy and science instruction: Perspectives on theory and practice,* ed. E. W. Saul, 13–32. Newark, International Reading Association.

Joyce, J. 1922. *Ulysses.* Repr., New York: Vintage, 1990. Page reference is to the 1990 edition.

Keeley, P., F. Eberle, and L. Farrin. 2005. *Uncovering student ideas in science, volume 1: 25 formative assessment probes.* Arlington, VA: NSTA Press.

Keeley, P., F. Eberle, and J. Tugel. 2007. *Uncovering student ideas in science, volume 2: 25 more formative assessment probes.* Arlington, VA: NSTA Press.

Keeley, P., F. Eberle, and C. Dorsey. 2008. *Uncovering student ideas in science, volume 3: Another 25 formative assessment probes.* Arlington, VA: NSTA Press.

Keeley, P., and J. Tugel. 2009. *Uncovering student ideas in science, volume 4: 25 new formative assessment probes.* Arlington, VA: NSTA Press.

Klentschy, M. 2005. Science notebook essentials: A guide to effective notebook components. *Science and Children* 43 (3):24-27.

Lee, O. 2005. Science education and student diversity: Summary of synthesis

and research agenda. *Journal of Education for Students Placed At Risk* 10 (4): 431–440.

Padilla M. J., K. D. Muth, and R. K. Padilla. 1991. Science and reading: Many process skills in common? In *Science learning: Processes and applications,* eds. C. M. Santa and D. E. Alvermann, 14–19. Newark, DE: International Reading Association.

Palincsar, A. S., and S. J. Magnusson. 2001. The interplay of firsthand and text-based investigations to model and suport the development of scientific knowledge and reasoning. In *Cognition and instruction: Twenty-five years of progress,* eds. S. Carver and D. Klahr, 151–194. Mahwah, NJ: Lawrence Erlbaum.

Postman, N. 1979. *Teaching as a conserving activity.* New York: Delacorte.

Reiss, M. J. 2002. Reforming school science education in the light of pupil views and the boundaries of science, *School Science Review* 84 (307).

Rosebery, A. S. and B. Warren, Eds. 2008. *Teaching Science to English Language Learners: Building on students' strengths.* Arlington, VA. NSTA Press.

Winnokur, J., and K. Worth. 2006. Talk in the science classroom: Looking at what students and teachers need to know and be able to do. In *Linking Science and Literacy in the K–8 classroom,* ed. R. Douglas, M. Klentschy, and K. Worth, 43–58. Arlington, VA: NSTA Press.

THE STORIES AND BACKGROUND MATERIAL FOR TEACHERS

EARTH SYSTEMS SCIENCE AND TECHNOLOGY

Core Concepts	The Coldest Time	Is the Earth Getting Heavier?	The Moon Around the World	Sunrise, Sunset
Solar energy	X			
Temperature	X			
Heat	X			
Radiational cooling	X			
Weather	X			
Climate	X			
Recycling of Matter		X		
Decay		X		
Decomposition		X		
Conservation of Matter		X		
Closed System	X	X		
Equinox				X
Solstice			X	X
Latitude			X	X
Earth's Tilt				X
Reflection			X	
Revolution			X	X
Moon Phases			X	
Earth-Moon-Sun System			X	X

CHAPTER 5
THE COLDEST TIME OF THE DAY

National Park Ranger Rudi was in charge of the program in the Everglades National Park called "Into the Wild." In this program, different families from the Miami area go camping for two nights. The families were picked from those who had never been to the Everglades or ever camped outdoors. It was not hard to find families that hadn't done either, but it was difficult to convince them that they should try out the experience. The kids—and even the adults—were afraid of the darkness and the wild animals in the area. Finally Rudi managed to find one family who was willing to go. The family had a mom, dad, and two children—a girl, Sasha, and a boy, Gene.

When they got to the park, Mom said that she hadn't gotten a wink of sleep the night before. She was worried about her children. Would they like the experience? Would they be safe? Would they learn about the Everglades and want to go back?

The family met Ranger Rudi in the parking lot of the visitor center. They did not own any of their own equipment, so Rudi was prepared with binoculars, food, cooking equipment, water bottles, sleeping bags, and tents. It was the dry season in January, but the nights could get cool, so Rudi had nice warm sleeping bags and jackets for everyone, since people in Miami were not used to cool weather and did not own much in the way of warm clothes.

After a full day of activities in the park including looking at birds, turtles, fish, and alligators, Rudi helped the family pitch their tents in the campground. There were lots of other visitors around so that the family felt safer surrounded by people. They had a campfire, toasted marshmallows and hot dogs, and soon it was time for bed. It got dark at about 5:30 p.m. and the Sun was due up the next morning at about 6:00 a.m.

"If it gets cold in the middle of the night, can I come and get in with you?" said Sasha, the littlest one, to her parents.

"Sure," said Dad, "As long as it is not really early in the morning."

"Oh, it won't be," said Gene, "'Cause it will be coldest around midnight."

"I'll bet it will be coldest about three in the morning," said Sasha. "And that *is* early in the morning, real early!"

"Well, when I'm talking about early in the morning, I mean about sunrise," said her father. "Don't come knocking at my tent when I'm just settling down for my pre-coffee sleep."

Rudi broke in and asked, "When do you all think it will be the coldest outdoors tonight, just after midnight, at three in the morning, or at sunrise?"

Sasha said three in the morning, Gene stuck to his prediction of midnight, and Dad was afraid it was going to be right around dawn. Mom was already in her bag and didn't have a clue. She just wanted to get the tent zipped up so no critters could get in.

Rudi said that it might be interesting to find out who was right, but how could they do it?

"Let's sleep on it," her father said, and they did. And the next morning, while they were eating pancakes cooked over a fire, they discussed how they could do it.

PURPOSE

The purpose of this story is to help students learn about the source of heat energy that warms their planet. Of course that is the Sun, and it only has an effect on the temperature of the Earth when it is shining on a particular spot. Another purpose is to stimulate the students to design a way to find out when, under normal conditions, the temperature in their area is lowest.

RELATED CONCEPTS

- Solar energy
- Temperature
- Heat
- Radiational cooling
- Weather
- Climate

DON'T BE SURPRISED

Your students may be unaware of the radiational cooling that takes place during the night when there are clear skies that allow heat energy to leave the Earth's surface. They may also be of the opinion that the "witching hour" of midnight has some hidden meaning or that sometime in the middle of the night the promise of sunshine will begin to warm the Earth. We have also found that few students realize that even after the Sun appears on the horizon it takes a while before the Sun is high enough in the sky to actually warm the Earth in any particular location.

CONTENT BACKGROUND

Most students and adults would agree that the Sun is the source of heat that warms our planet. We hear enough about radiational cooling from the TV weather reports to know that nights with clear skies allow the heat energy of the Earth to radiate out into space, unimpeded by insulating cloud layers. This simple law of thermodynamics is so much a part of our lives that the idea of heat leaving a warm place and moving to a cooler place is usually not a problem. No, the problem, as shown in the story, is predicting when it is that the heat starts returning. Since the glow of the sunrise signals the appearance of our heat source, it is reasonable to assume that a rise in temperature is expected immediately. But, anyone who has been outdoors at sunrise has felt the cold air and experienced the delay of the eventual warming. It is difficult to understand that more direct rays of the Sun are needed to produce any discernable difference in air temperature.

The Sun's rays have to actually warm up the surface of the Earth, so they must be at an angle that is sufficiently high to make a difference. An analysis of the hourly tem-

perature increases show that the temperature rises slowly in the early morning hours and often does not reach highs for the day until late afternoon. This means that as the day progresses, and if the Sun continues to shine, the Earth absorbs more heat than it radiates back so that it warms up. This implies that the Earth is continuously receiving heat energy and is also losing it at the same time. Due to the high angle of the Sun in the afternoon hours, the incoming heat exceeds the outgoing heat, sometimes even after sundown, when the Earth may lose heat slowly due to cloud cover.

For those who live in a climate where winter frost leaves a coating on the ground, there is a dramatic source of evidence for radiational heating. On the white frosting of ice crystals on the ground, we can see shadows of trees, posts, and other objects. As the sun rises higher in the sky, we see something I call "frost shadows," because frost begins to melt *except* where there is a shadow that blocks the Sun's rays from striking the ground. So these white "shadows" are caused by the lack of radiation from the Sun. As the Sun rises higher and higher, the white shadows slowly disappear as the angle and direction of the Sun changes. This shows that the Sun's rays are responsible for the warming of the Earth's surface, eventually melting the frost. It also demonstrates, indirectly, that the temperature of the Earth's surface takes a time to rise, which is why the frost hangs around for a while. And it provides evidence that the lower angles of the Sun's rays are not as strong as the higher angles.

This phenomenon also shows that the Sun's position in the sky changes not only in its rise above the horizon, but toward the south and eventually the west. Thus, the white shadows begin to disappear in a clockwise direction as the Sun changes position. The wider the shadow, the easier it is to see this happen and perhaps even measure the Sun's movement, both in direction and speed. Watching the frost shadows disappear from a large tree's shadow takes more time that from a thin pole or sapling. For teachers, the timing of the phenomenon of frost shadows is great. By the time the Sun rises high enough to peek over the roof of the school, the frost is still on the ground. Classes can observe the frost during the opening hours of the school day and have plenty of time for discussion afterward.

During this discussion, someone usually beings up the topic of climate. A student may say, "Climate is what you expect, weather is what you get." Climate, however, is determined by many variables such as altitude, latitude, proximity to water, wind, rainfall, and other conditions that are examined and recorded over a period of at least 30 years. Thus we have deserts, rain-forests, cloud forests, tropics, tundras, and so on. For example, south Florida is listed as a subtropical climate, even though other places around the world at the same latitude are deserts. What makes its climate different than the Sonoran, Gobi, or Sahara deserts is that south Florida is surrounded by water and warm currents that combine to provide it with a wet season and a dry season. It is classified as subtropical because the usual definition of *tropical* entails that the latitude is within 23.5° north or south of the equator. South Florida is about 25° north of the equator, so, along with its unusual tropical weather, it can get frost and even extended freezes, like the one just experienced during the winter of 2009–10. These are rare, occurring only once or twice a century, however.

In the context of this story, climate might have a distinct effect upon the temperature range seen in one day. For instance, in a desert, the temperature might range from freezing (below 32° Fahrenheit) at night to the low hundreds during the day. This would correspond with the situation described in the story with the lowest temperatures being around dawn. But, if this area were surrounded by mountains that block the Sun's rays, it would delay the rise in temperature until the Sun was at an altitude higher than the surrounding mountains.

Radiation or *radiative cooling* is the process whereby the Earth loses heat by emitting long-wave (*infrared*) radiation, which balances the short-wave or *visible* energy from the Sun. The cooling of the Earth is very complex, having to do with transference of heat due to convection (air currents), evaporation, and other things, such as geographical factors. Radiative cooling goes on all the time, but is especially intense under conditions such as nighttime clear skies, light winds, and low humidity.

related ideas from the national science education standards (Nrc)

K–4 Objects in the Sky
- The Sun provides the light and heat necessary to maintain the temperature of the Earth.

5–8 Transfer of Energy
- The Sun is a major source of energy for changes on Earth's surface. The Sun loses energy by emitting light. A tiny fraction of that light reaches the Earth, transferring energy from the Sun to the Earth. The Sun's energy arrives as light with a range of wavelengths consisting of visible light, infrared radiation, and ultraviolet radiation.

5–8 Earth in the Solar System
- The Sun is the major source of energy for phenomena on Earth's surface.

related ideas from benchmarks for science literacy (aaas)

K–2 Energy Transformation
- The Sun warms the land, air, and water.

3–5 The Earth
- The weather is always changing and can be described by measurable quantities such as temperature, wind direction and speed, and precipitation.

6–8 The Earth
- The temperature of a place on Earth's surface tends to rise and fall in a somewhat predictable pattern every day and over the course of a year.

6–8 Energy Transformation
- Heat can be transferred through materials by the collisions of atoms or across space by radiation.

USING THE STORIES WITH GRADES K–4

One of the first things that would benefit younger children would be to become familiar with the thermometer—how to read one and the range of a typical thermometer. Many teachers prefer to begin with a thermal strip thermometer, the kind that is used on the forehead to take bodily temperatures. Prepare several containers of water that are of various temperatures, yet within the comfort and safety zone. Children immerse their hands in the water and then compare the sensation with the reading on the thermometer. The concepts of *cool* and *warm* may then take on a numerical value. They may realize that the thermometer is a better indicator of temperature change than their senses.

To get an idea of the power of the solar energy that comes from the Sun, you may want to purchase some of the least expensive construction paper, that which will most likely fade in sunlight. If weather permits, place this paper on the ground or on a table in direct sunlight. Have the students place some objects such a saucers or blocks that will block the sunlight from reaching the paper. These can even be done in an artistic manner (to include art in the science lesson). After the Sun has had a few hours to fade the paper, the images will be there to see. This can lead to an investigation with many questions, including:

- Will the Sun create the images if the activity is done inside on a sunny windowsill through glass?
- Will afternoon sunlight work more quickly than morning sunlight?
- Will afternoon sunlight make darker images than early morning sunlight?
- Will sunshine in different times of the year cause different results?
- How do the differences in images change as the length of your shadow changes?

Making "thermosicles" (thermometers immersed in ice cubes) can also be interesting for students to observe. These are made by putting thermometers in an ice cube tray so that the bulbs are in the water, and the register can act as a handle. This will show the temperature of the ice, and it will remain at 0°C even as the ice begins to melt. Thus the child can see the temperature and feel the ice cube at the same time.

USING THE STORIES WITH GRADES 5–8

Children of this age group may well be more adept at using the internet or public press to gather information about the topic addressed in this story. The internet is useful because it is very unlikely that a student will volunteer to take temperature readings at each hour during the day and night (and unreasonable to ask!). Students can keep a record of a few weeks of hour-by-hour temperature cycles in their science notebooks so that they can draw some conclusions from the patterns they see over time.

You may find that giving the probe, "Camping Trip" in *Uncovering Student Ideas in Science, volume 4*, along with the story as a discussion starter will help your students become more fully immersed in the problem (Keeley and Tugel 2009). The probe may also be used as a formative assessment tool or even as a follow-up tool after the topic has ended.

Children who live above the latitude of the tropics will be familiar with seasonal changes in outdoor temperatures but will still be surprised that the lowest temperatures occur just about dawn. An activity with flashlight and paper can be helpful here. The student holds a flashlight directly over a piece of paper in a darkened room and circles the area that is lit by the beam, which represents the Sun's rays at midday or early afternoon. The flashlight then is shown upon the paper at an angle, (representing dawn) and the lightened area is again circled with a pencil or pen. The area of the light shown at an angle is larger than the area produced by the overhead position of the flashlight. This can be explained to represent the thermal energy received from the Sun as being spread over a larger area and therefore less potent at any given spot within the lighted portion. Heat lamps can also be used with thermometers in the same way to show that temperatures will rise less when the lamp is held at a lower angle than when shone directly down on the thermometer.

When the data are analyzed and the temperatures at dawn are seen to be the lowest, on average, the story can be finished. Dad probably did have someone "knocking on his tent" early in the morning by youngsters who were experiencing the coolest temperatures of the day.

Should the topic of climate come up, information given above might be of value to you. It is not necessarily within the realm of this particular story but the ramifications of climate and climate change are often of interest to students.

related NSTa Press Books and NSTa Journal articles

Keeley, P., 2005. *Science curriculum topic study: Bridging the gap between standards and practice.* Thousand Oaks, CA: Corwin Press.

Keeley, P., F. Eberle, and J. Tugel. 2007. *Uncovering student ideas in science: 25 more formative assessment probes, vol. 2.* Arlington, VA: NSTA Press.

Keeley, P., F. Eberle, and C. Dorsey. 2008. *Uncovering student ideas in science: Another 25 formative assessment probes, vol. 3.* Arlington, VA: NSTA Press.

Keeley, P., and J. Tugel. 2009. *Uncovering student ideas in science: 25 new formative assessment probes, vol. 4.* Arlington, VA: NSTA Press.

Konicek-Moran, R. 2008. *Everyday science mysteries: Stories for inquiry-based science teaching.* Arlington, VA: NSTA Press

Konicek-Moran, R. 2009. *More everyday science mysteries: Stories for inquiry-based science teaching.* Arlington, VA: NSTA Press.

Konicek-Moran, R. 2010. *Even More Everyday Science Mysteries: Stories for inquiry-based science teaching.* Arlington, VA: NSTA Press.

references

American Association for the Advancement of Science (AAAS).1993. *Benchmarks for science literacy.* New York: Oxford University Press.

Childs, G. 2007. A solar energy cycle. *Science and Children* 44 (7): 26–29.

Diamonte, K. 2005. Science shorts: Heating up, cooling down. *Science and Children.* 42 (9): 47–48.

Driver, R., A. Squires, P. Rushworth, and V. Wood-Robinson. 1994. *Making sense of secondary science: Research into children's ideas.* London and New York: Routledge Falmer.

Gilbert, S. W., and S. W. Ireton. 2003. *Understanding models in earth and space science.* Arlington, VA: NSTA Press.

Keeley, P., and J. Tugel. 2009. *Uncovering student ideas in science: 25 new formative assessment probes, vol. 4.* Arlington, VA: NSTA Press.

Konicek-Moran, R. 2008. *Everyday science mysteries.* Arlington, VA: NSTA Press.

Konicek-Moran, R. 2009. *More everyday science mysteries.* Arlington, VA: NSTA Press.

National Research Council (NRC). 1996. *National science education standards.* Washington, DC: National Academy Press.

Oates-Brockenstedt, C., and M. Oates. 2008. Earth science success: 50 lesson plans for grades 6–9. Arlington, VA: NSTA Press.

CHAPTER 6
IS THE EARTH GETTING HEAVIER?

Tom looked up from raking leaves and said to his cousin, Laura, "I think the Earth is getting heavier. Look at all of these leaves around us. There are more each year, and they weigh something, don't they?"

Laura leaned on her rake and looked at Tom with doubt. "Where did you get that brilliant idea?" she asked. "Is it because you hate raking so much that you think we just ought to let them sit there and add more weight to the Earth?"

"Well, it makes sense doesn't it? Each year the tree makes lots of leaves and each fall, they fall down and lie on the ground. Look over there in the woods; there are tons of leaves from last year and underneath those leaves are the ones from the year before. They have to add up to something."

"Well, I guess so and this has been going on for millions of years so the old planet must be getting really fat by now. No wonder the scientists say the Earth is slowing down," said Laura with a smile on her face.

I guess she doesn't really believe me, thought Tom.

"Okay," he said, "Let's go into the woods and take a look at the leaves on the ground and I'll show you what I mean."

The two children walked over to the woods and began sifting through the layers of leaves.

"See," said Tom, "Here are this year's leaves and just below are last year's leaves and below them are the leaves from the year before! I admit that they are a bit soggy and beat-up looking but they are still there."

"Okay," said Laura," But where are the leaves from three years ago and four years ago?"

"I think they are under the ground somewhere," said Tom. "Let's dig some up and see what happened to them."

And so they did and what they found settled the argument.

PURPOSE

My research and that of others show that children have a difficult time understanding the recycling of organic matter in an ecosystem. This story aims to have students speculate about what happens to organic material over time.

RELATED CONCEPTS

- Recycling of matter
- Decay
- Decomposition
- Conservation of matter
- Closed system
- Open system

DON'T BE SURPRISED

Unless students have some understanding of the particulate nature of matter, it is difficult if not impossible for them to understand how matter can be broken down into parts so small that they can be recombined with other matter to form new compounds and impact new living things. Your students may actually believe that leaves and wood from fallen trees over the millennia have accumulated to the extent that they actually increase the mass of the Earth. After all, don't we see little saplings grow into giant trees? Doesn't that add mass to the Earth?

CONTENT BACKGROUND

Obviously, we do not want our students to think that the Earth is accumulating leaves over the millennia and growing more obese. Nor do we want them believing that growing trees and other plants add mass to the Earth each year. It is technically possible that meteorites add a miniscule amount of weight when they enter Earth's atmosphere and finally land on the Earth's surface. True, some hydrogen atoms are lost into space, and rockets and satellites leave our planet for their homes in space, but these are almost negligible. Even though students may bring it up, the amount of mass lost or gained is not worth considering.

We are talking in this story about many, many tons of leaves that have fallen since plants and trees evolved. If you do any walking in the woods in a climate where deciduous trees lose their leaves, you are aware of the depth that these leaves can accumulate. If you have a yard surrounded by trees, the raking and mulching of these leaves takes up a great deal of your effort and time. Yes, the amount of leaves that fall to the ground from trees is substantial. If they stayed in place, I think you can imagine how tall the pile would be by now.

Digging beneath the pile of this year's leaves, you will come across some rather nasty looking wet and bedraggled looking pieces of leaves, partial ghosts of their former selves. They have already begun the process of being devoured by decomposers and broken down into their smallest parts, the molecules that made them working leaves, former producers of sugars and starches. There will be numerous animals—mainly invertebrates like earthworms, millipedes, pill bugs—and many varieties of bacteria and fungi at work breaking down the original leaves and plant material into compost, a rich organic soil. They do this by digesting the substance of the organic material to release the building blocks of living organisms into the soil. The further down you are able to go into the leaf litter, the less the leaves look like leaves and the more they look like soil. These decomposers do a good job, leaving little that is recognizable. Then other living things, including the trees themselves, absorb those building blocks so that they might live and grow. The cycle is complete.

Most of my research and that of my colleagues on the concept of decomposition has shown that children and adults alike believe that dead materials turn directly into soil, as if by magic (Leach, Konicek, and Shapiro 1992; Leach et al. 1992; Sequeira and Freitas 1986. Exceptions to this were students who had some understanding of the particulate nature of matter and believed that it was made up of tiny particles capable of being reused by other living things.

How can all of these trees and plants continue to grow and not add to the mass of the Earth? The answer may be even less believable for your students. Earth is considered a *closed system* since the materials on the planet are finite and stay within the confines of the planet and its atmosphere. As I mentioned previously, virtually no *mass* is exchanged with the other parts of the universe, but of course energy is, particularly the radiant energy from the sun. When plants make the food and the substances that define their structure and existence, they use carbon dioxide from the atmosphere as their source of carbon. The process of photosynthesis relies on the carbon dioxide in the air and the energy of the Sun to produce the matter we know as living botanical organisms—plants.

But many people believe that the air around us has no weight and no substance. So how can something that has no mass be the source of the carbon that makes up the mass of every living plant? This is a difficult concept for many to understand unless they are convinced otherwise.

When plants decay, the materials are returned to the planet in molecular form, suitable for reuse by other growing things. Thus the mass is conserved over time, so the mass does not change. (See the probe "Seedlings in a Jar," which suggests that you place seedlings or seeds in a sealed jar of soil and let the seeds germinate and grow. If the total system is weighed before and after the growth period, it will weigh the same. No matter left or entered the jar so it is a closed system just like the Earth (Keeley, Eberle, and Farrin 2005). You may want to try this yourself if you are in need of convincing that the closed system will be entirely self-sufficient. You may also find some of the same concepts in the story "Springtime in the Greenhouse" in this book (p. 103).

If you have a copy of *More Everyday Science Mysteries*, read chapter 7, "Rotten Apples," for another take on the decomposition process (Konicek-Moran 2009).

related ideas from the national science education standards (NRC)

K–4: *The Characteristics of Organisms*

- Organisms have basic needs.

K–4: *Organisms and their Environments*

- All organisms cause changes in the environment where they live. Some of these changes are detrimental to the organism or to other organisms, whereas others are beneficial.

5–8: *Populations and Ecosystems*

- Decomposers, primarily bacteria and fungi, are consumers that use waste materials and dead organisms for food.

related ideas from benchmarks for science literacy (AAAS)

K–2: *Flow of Matter and Energy*

- Many materials can be recycled and used again, sometimes to different forms.

K–2: *Constancy and Change*

- Things change in some ways and stay the same in some ways.

3–5: *Interdependence of Life*

- Insects and various other organisms depend on deap plant and animal material for food.
- Most microorganisms do not cause disease and many are beneficial

3–5: *Flow of Matter and Energy*

- Some source of energy is needed for all organisms to stay alive and grow.
- Over the whole Earth, organisms are growing, dying, and decaying, and new organisms are being produced by the old ones.

Using the Stories with Grades K–4

I suggest that you give the probe "Rotting Apple" from *Uncovering student ideas in science, volume 3: Another 25 formative assessment probes* (Keeley, Eberle, and Dorsey 2008). This will give you an idea of what your students already think about decomposition. Unless I miss my guess, I'll bet that most of your students will choose the option that says wind and water soften the apple and it just dissolves into soil.

With students of this age, it is appropriate for them to observe an apple or some other fruit decompose in a covered container. Making sure the container inside the classroom is closed assures that the fungal spores stay within the container and do not contaminate the room. The presence of obvious fungi will give your students a visual appreciation of their work in decomposing the fruit.. In warm weather, it's a good idea to place a container with holes in its lid on the outside windowsill so that it can be observed but not touched. This allows flies to visit the fruit and deposit eggs that hatch and begin to devour the rotting fruit. If maggots or grubs are present, it will give students the idea that larger animals also play a part in decomposition.

Fungi are considered a separate kingdom in the natural world classification order. They get their nutrition by decomposing dead organisms and absorbing the nutrients they need to live and grow. They reproduce by spores that are always present in our air. These spores are the progenitors of the fungi that take advantage of any food left out and exposed to them. They can also be dangerous to some people who are allergic to them. Mold, for example, is a fungus that causes allergic reactions in many people. Entire buildings have been closed because of mold contamination and even condemned and destroyed.

If you do not want to have a decomposing chamber in your classroom, you may avail yourself of the materials on the internet. To view a time-lapse video of fruit and vegetables decomposing, check out *www.metatube.com/en/videos/27174/Fruit-and-Vegetable-Decomposition-Time-lapse*. Some people find decomposition gross and I think you should view it first before showing it to young children. My personal experience is that viewing how fruit and vegetables decompose through time-lapse technology amazes even young children. Of course, middle school children will find it even more intriguing. You will also notice that the sprouting and growing of certain of the vegetables in the compost being formed shows the cycle of life in graphic form.

You may want to add a chart of students' "best thinking" on what happens to leaves and other things that fall to the ground. If this is done before letting them look through leaf litter or watching the video of the decomposing materials, they can modify their findings through their experience. (Caution: Leaf litter is a favorite habitat for scorpions, centipedes, and coral snakes if you live in a tropical or subtropical climate. It is best to have your students work with small rakes to find the lower levels of leaves so that they can be a safe distance if they come upon one of these poisonous critters.)

The story in this chapter also goes quite well with the stories and related activities in "Rotten Apples" and "Worms are for More than Bait," both found in *More Everyday Science Mysteries* (Konicek-Moran 2009). All three stories focus on what I consider one of the most important biological concepts—one that is one of the least understood by children and adults alike—decomposition. Understanding that composting (recycling organic material) lessens the burden on landfills promotes stewardship of the Earth. Much of the stuff that is put into our landfills is organic material that could be recycled into usable material. You will find a great deal of information on the Environmental Protection Agency website about what kinds of things can and cannot be composted (*www.epa.gov/epawaste/conserve/rrr/composting/index.htm*).

USING THE STORIES WITH GRADES 5–8

Middle school students are more likely to agree with Laura in this story. Giving the probe mentioned in the K–4 section will give you a good look at prior conceptions. Be sure to pay close attention to students' rationalizations in the written section of the probe. It is often in this that you can see if there is real understanding. Only Selma's answer, "I think small organisms use it for energy and building materials," mentions the decomposers. Your students' written explanations will confirm for you whether they really understand the role of decomposers in the process.

If your class ventures outside to recreate the actions of the children in the story, they will certainly notice the differences in the levels of leaves on the ground. (Note the caution about leaf litter containing some dangerous insects and snakes in warmer climates that I made above). If students are allowed to dig even further beneath the soil they will come upon the macro animals like worms, pill bugs, millipedes and others that begin the process of decomposing the leaves and other organic material. Help your students to notice the fungi that are also covering the leaves. Looking around they will notice that the dead trees, either standing or lying on the ground will be covered with fungi. The trees, both standing and fallen often are covered with shelf fungi commonly called "turkey tails" (*Trametes elegans*). These fungi and other plants that grow on the fallen logs help break down the dead trees. Often in moist seasons, trees will take on the role of "nurse logs," as the decomposition of their upper surfaces and the addition of materials falling from the trees above will form a growing platform. They are called nurse logs since they host a great many plants that continue the process of decomposition. These plants are often ferns, but in certain climates, can actually be much larger plants such as trees. The roots of these plants take in the decomposed materials from the nurse log, and thus help to reduce it even further by reusing it to nurture their own growth.

Now let's tackle the concept of plants growing in a closed system and help students understand that plants take carbon from the air in order to make food and build their structure. Give the probe mentioned earlier, "Seedlings in a Jar." This time, use the probe for a discussion starter. Let the students debate, making sure that they give their reasons why they think the system in the jar will weigh more,

less, or the same after the seeds germinate and grow. Then let the students design and set up an investigation using the materials mentioned in the probe. This is the topic featured in the chapter of the same name, "Seedlings in a Jar," in this book. However, be advised that the chapter in this book focuses more on systems. Set up properly, the system will not change weight at any time. Make sure that variables are fully identified and that the investigation is completely fair. It would be a good idea to weigh each part of the system separately: seeds, earth, water, jar, and lid, in order to have these data during the discussion afterward.

Expect that the students might:

- say that the mass for producing the growth of the seedlings came from the soil;
- say that the mass for producing the growth of the seedlings came from the water; or
- accuse someone of adding soil or water or removing some.

Sooner or later the arguments will convince them that whatever happened within the jar used only the matter that was there in the first place. This may not persuade all of your students that the atmosphere contributed the material for the plants to germinate and grow. (The plants will weigh more afterward, after all, and the soil will have not changed significantly, at least not enough to account for the growth of the plants in the jar.) However, it will be another plank in the scaffolding that will sooner or later help them to understand that the atmosphere, indeed, does provide the material that builds the mass of growing plants: carbon from the carbon dioxide.

A similar type of investigation can be carried out in a closed system using decomposers such as worms, bugs, or fungi. An important part of these investigations is that students have the opportunity to plan and carry out investigations and consider the variables involved. Do not use sterilized soil, as it contains no living organisms. You can use leaves or fruit or vegetables to compost. There will be sufficient spores in most dirt to produce fungi. Just remember that if you use animals such as worms or bugs, keep the transparent parts of the jar covered so that the light will not disturb them. The covering can be removed for short intervals in order to allow observation and drawing of what is happening over time.

related NSTa Press BOOKS and NSTa JOURNAL articles

Keeley, P. 2005. *Science curriculum topic study: Bridging the gap between standards and practice.* Thousand Oaks, CA: Corwin Press.

Keeley, P., F. Eberle, and L. Farrin. 2005. *Uncovering student ideas in science, volume 1: 25 formative assessment probes.* Arlington, VA: NSTA Press.

Keeley, P., F. Eberle, and J. Tugel. 2007. *Uncovering student ideas in science, volume 2: 25 more formative assessment probes.* Arlington, VA: NSTA Press.

Keeley, P., F. Eberle, and C. Dorsey. 2008. *Uncovering student ideas in science, volume 3: Another 25 formative assessment probes.* Arlington, VA: NSTA Press.

Keeley, P., and J. Tugel. 2009. *Uncovering student ideas in science, volume 4: 25 new formative assessment probes.* Arlington, VA: NSTA Press.

Konicek-Moran, R. 2008. *Everyday science mysteries: Stories for inquiry-based science teaching.* Arlington, VA: NSTA Press.

Konicek-Moran, R. 2010. *Even More Everyday Science Mysteries: Stories for inquiry-based science teaching.* Arlington, VA: NSTA Press.

references

Driver, R., A. Squires, P. Rushworth, and V. Wood-Robinson. 1994. *Making sense of secondary science: Research into children's ideas.* New York: Routledge Falmer.

Keeley, P., F. Eberle, and C. Dorsey. 2008. *Uncovering student ideas in science, volume 3: Another 25 formative assessment probes.* Arlington, VA: NSTA Press.

Leach, J., R. Konicek, and B. Shapiro. 1992. The ideas used by British and North American school children to interpret the phenomenon of decay: A cross-cultural study. Paper presented to the annual Meeting of the American Educational Research Association, San Francisco.

Leach, J., R. Driver, P. Scott, and C. Wood-Robinson. 1992. *Progression in conceptual understanding of ecological concepts by pupils aged 5–16.* Leeds, UK: The University of Leeds, Centre for Studies in Science and Mathematics Education.

Time lapse video of decomposition. *www.metatube.com/en/videos/27174/Fruit-and-Vegetable-Decomposition-Time-lapse/*

Konicek-Moran, R. 2009. Rotten apples. *In More everyday science mysteries: Stories for inquiry-based science teaching.* 61–68. Arlington, VA: NSTA Press.

Konicek-Moran, R. 2009. Worms are for more than bait. *In More everyday science mysteries: Stories for inquiry-based science teaching,* 91–100. Arlington, VA: NSTA Press.

National Research Council (NRC). 1996. *National science education standards.* Washington, DC: National Academy Press.

Sequeira, M., and M. Freitas. 1986. Death and decomposition of living organisms: Children's alternative frameworks. Paper presented at the 11th Conference of the Association for Teacher Education in Europe (ATEE), Toulouse, France.

WHAT'S THE MOON LIKE AROUND THE WORLD?

Aaqil and Durrah had recently arrived in the United States from the Middle East and were enjoying the night sky in their new home in the 'Middle West.' The light from the Sun had disappeared, and the crescent shaped Moon had just come into view low in the western sky. Aaqil began to wonder out loud.

"I think that the Moon is very beautiful, Durrah. I love this shape that looks like a curved circle. I wonder what the Moon looks like tonight in our home country?"

"Yes, it looks very beautiful, and I wonder, too, what it looks like way down south in Australia. They say everything is backward and upside down in the country they call 'Down Under,'" she replied.

"I think," said Aaqil, "They call it 'Down Under' because it is way south of the equator. Our teacher says that compared to us, people down there are standing upside down. I suppose they see everything like they were standing on their head!"

"So, if we have a full Moon here, they would have a new Moon and see nothing in the sky, right?"

"I'm not so sure about that, Durrah," said her brother. "I think maybe the moon is the same everywhere on Earth, but I don't really know."

"Well, our country across the ocean is a long way from here, even though we are both in the northern part of the world. So how can the Moon be the same shape?"

"It is the same Moon, isn't it?"

"Yes, but it is so far away from what we are seeing that it can't be the same shape!"

"There must be a way to find out."

"I know! We can email our friends we left behind. They would answer our question about what the Moon looks like from east to west," said Durrah. "But what about north to south?"

"Let's try to find somebody who lives way down south."

PURPOSE

A great deal of confusion arises from the lack of understanding about the Moon's journey around the Earth and its apparent shape changes. This investigation is aimed at confronting this confusion by looking at the Earth-Moon system and how it appears from various points on the Earth. Since the discussion includes the two hemispheres, children will be able to see that all of the phases of the Moon are opposites in every way.

related CONCEPTS

- Reflection
- Orbits
- Earth-moon-sun relationship
- Systems
- Moon phases

DON'T BE SURPRISED

The causes of Moon phases are an enigma to many children, as well as adults. Witness the responses given by the graduates of Harvard University and students in a nearby high school in the film "Private Universe" (Schneps 1987). Things haven't changed a lot, from my experience, in the last 24 years and it is likely that your students will have an interesting time discussing and arguing the points made in the story. Many students and adults believe that the shape of the moon as it changes phases is due to the shadow of the Earth on the Moon. It is indeed one of the strongest astronomical misconceptions.

CONTENT BACKGROUND

Imagine a situation where the Moon's phases are different all across the world. Now imagine you have to prepare a calendar for sale around the world with these Moon phases pictured each month. What a mess! It's a good thing this is not the case. The Moon looks pretty much the same no matter where you live *unless* you live in the southern hemisphere. If you live far enough "down under" you are literally standing on your head compared to those who live in the northern hemisphere, so you must see the shape of the Moon differently. However, anywhere on the Earth we see the sunlit portion of the Moon, and depending upon our position in regard to the Earth and Moon, that sunlit portion of the Moon will appear different each day. This means that it is impossible for some of us to see a full Moon and others to see a new Moon.

If you live in the United States and see a full Moon, what do you expect a friend living in Chile, South America to see? In volume 1 of Page Keeley's *Uncovering Student Ideas in Science* (2005) there is a probe called "Gazing at the Moon"

that asks this very question. Many people believe that because of the hemispheric difference, the difference in Moon phases is opposite. This is true, but it does not mean that a full Moon in the northern hemisphere becomes a new Moon in the southern hemisphere. Rather, it means that the crescent shape is pointing in the opposite direction.

This difference will challenge your spatial relationships to their limits. Imagine that you are looking at the Moon in its crescent phase. Now, imagine that you are looking at the same image while standing on your head. The horns of the crescent would be pointing in the opposite direction, right? Try it and see. If you can't stand on your head just bend over until you are looking upside down.

The Earth rotates in a west-to-east direction so the Moon will rise in the east and set in the west regardless of where you live. However, if you live in the southern hemisphere (SH) the Moon will appear to travel across the northern sky rather than in the southern sky as it does in the northern hemisphere (NH). In the NH, the lighted portion of the Moon grows from right to left over the 29.5 days of its full cycle. In the SH the lighted portion of the Moon grows from left to right.

Because of this, the rule we have about the waxing and waning Moon is opposite in the NH and SH. In the NH we say that the Moon is waxing (moving toward full phase) if the shape of the first quarter looks like a growing C and the last quarter looks like a D. The opposite is true in the SH. I found that I needed some help on this so I used a website to visualize the differences. One good one is *www.woodlands-junior.kent.sch.uk/time/moon/phases.html* (Barrow 2008). Scroll down to the bottom and click on "Moon Around the World." It also mentions that near the equator, the crescent moon looks either like a U or a smile when it sets. This is because the Moon's orbit is very close to that of Earth's orbit and so is at less of an angle than it is in either the far NH or SH.

For those of you who are beginning to zone out and say this is much too difficult for your students, I recommend you read "Meeting the Moon From a Global Perspective" (Smith 2003). In this article, the author chronicles a project that connected students from all parts of the globe exchanging their views of the moon and how it related to their global position and their cultures. I wish I could have been part of that project and posit that, with the suggestions and caveats listed in the article, it is possible to re-create much of the same excitement and learning on your own. There are privacy concerns involved with children corresponding with each other, but these problems are addressed in the article. The gains in cultural understanding and science knowledge far overshadowed the difficulties in setting this project up. In fact, if you use your favorite search engine and type in "current global school science connections," you may be able to find some ongoing projects that may interest you and your school system.

RELATED IDEAS FROM THE NATIONAL SCIENCE EDUCATION STANDARDS (NRC 1996)

K–4 Changes in the Earth and Sky
- Objects in the sky have patterns of movement. The observable shape of the Moon changes from day to day in a cycle that lasts about a month.

5–8 Earth in the Solar System
- Most objects in the solar system are in regular and predictable motion. Those motions explain such phenomena as the day, the year, phases of the Moon and eclipses.

RELATED IDEAS FROM BENCHMARKS FOR SCIENCE LITERACY (AAAS 1993)

K–2: The Universe
- The Moon looks a little different every day, but looks the same again about every four weeks.

3–5: The Universe
- The Earth is one of several planets that orbit the Sun, and the Moon orbits around the Earth.

6–8 The Earth
- The Moon's orbit around the Earth once in about 28 days changes what part of the Moon is lighted by the Sun and how much of that part can be seen from the Earth—the phases of the Moon.

USING THE STORIES WITH GRADES K–4

Using this story with younger children will raise more questions about the phase changes in the Moon than anything else. Children with developing spatial skills will have a difficult time imagining people "standing upside down," in the southern hemisphere. The story, "Moon Tricks," in *Everyday Science Mysteries* (Konicek-Moran 2008) tells the story of a boy who moves into a new house and sees the full Moon in his bedroom window on his first night. He is disappointed on the second night to find the full Moon missing and awakes in the middle of the night a week later to see the Moon outlined in his window but in a different shape. The story

goes on to lead the students to observe the Moon over a period of time to see for themselves the pattern of phases that occur in the sky. Materials in Everyday Science Mysteries give elaborate directions for helping students keep Moon journals and use them to build models of the Earth-Moon system.

Even if you do not have access to this book, you can make arrangements with parents to help your students keep track of the Moon's direction, shape, and height above the horizon at a given time each day. This will result in a journal that can be discussed in class and give your students some firsthand data to analyze. Our experience is that children are not aware of these changes and will benefit greatly by observing them on a regular basis and collecting data about what they see.

USING THE STORIES WITH GRADES 5-8

If your students in the middle grades are not aware of the reasons for the Moon's apparent changes in shape over the monthly period, I suggest the same story and background material offered to grades K–4 teachers. It is important that your students understand the celestial mechanisms that cause the pattern of the phases of the Moon before they can visualize the way others in the world see the Moon.

Your students should work in pairs in a darkened room with a single lamp about two meters above the floor acting as the Sun. One student will act as the Moon using a Styrofoam ball on a dowel. The other student is the Earth and observes the Moon as it moves around this student slowly. The Earth rotates once (the child spins once, back to the original place) and the Moon moves in the same direction so that the phases are seen as they go through the month of 29 days. Partners switch and the exercise is repeated.

Once the students have seen how this works, they should be able to repeat the activity and try to visualize the phase changes by bending over, assuming a modified upside-down position and seeing that the pattern is reversed. It may take several tries before the reverse pattern is seen by all of the students. Have the students acting as the Earth vary their positions so that they can see that they must be very far down in the SH in order to see the reverse pattern. At this point, they should also be asked if they think the Moon would look any different to an observer in the same hemisphere.

They should notice that the lighted area of the Moon is always the same regardless of the position of the observer, but that the observer's view is dependent upon the rotation of the Earth. Whether in France, Asia, or anywhere else on the globe, the Earth has to turn sufficiently so that Moon is visible, and of course since the Earth rotates from west to east, the further west one lives, the later the Moon becomes visible. However, the lighted surface of the Moon is always the same shape regardless of hemisphere (i.e., crescent, quarter moon, full, and new). So Aaqil and Durrah in the story can be certain that their friends back home will be seeing the same moon shape that they do but their friends will have seen it earlier. In fact if they could call their friends, they would find that the Moon had come into view about seven hours earlier over there.

So, in essence, the Moon rises later each day but looks basically the same if all viewers are in the same hemisphere. The major change in views depends on being in different hemispheres. The SH and NH view the moon with a difference of 180°.

related NSTA Press Books and NSTA Journal articles

Gilbert, S. W., and S. W. Ireton. 2003. *Understanding models in earth and space science*. Arlington, VA: NSTA Press.

Keeley, P., and J. Tugel. 2009. *Uncovering student ideas in science, volume 4: 25 new formative assessment probes*. Arlington, VA: NSTA Press.

Konicek-Moran, R. 2009. *More everyday science mysteries: Stories for inquiry-based science teaching*. Arlington, VA: NSTA Press.

Oates-Brockenstedt, C., and M. Oates. 2008. *Earth science success: 50 lesson plans for grades 6–9*. Arlington, VA: NSTA Press.

references

American Association for the Advancement of Science (AAAS).1993. *Benchmarks for science literacy*. New York: Oxford University Press.

Barrow, M. 2008. The phases of the moon. Woodlands Junior School, Kent. UK. *www.woodlands-junior.kent.sch.uk/time/moon/phases.html*

Keeley, P., F. Eberle, and L. Farrin. 2005. Gazing at the moon. In *Uncovering student ideas in science, volume 1: 25 formative assessment probes*, 177–181. Arlington, VA: NSTA Press.

Konicek-Moran, R. 2008. *Everyday science mysteries: Stories for inquiry-based science teaching*. Arlington, VA: NSTA Press.

National Research Council (NRC). 1996. *National science education standards*. Washington, DC: National Academy Press.

Schneps, M. 1986. *The Private Universe Project*. Harvard Smithsonian Center for Astrophysics.

Smith, W. 2003. Meeting the moon from a global perspective. *Science Scope* 36 (8): 24–28.

CHAPTER 8
SUNRISE, SUNSET

"If you are lost in the woods, and it is not a cloudy morning, how can you tell your directions without a compass?" asked the scoutmaster during the weekly meeting of a Boy Scout troop in northern Idaho.

"Moss only grows on the north side of trees," responded from one of the troop members.

"Is that really always true?" probed the scoutmaster.

There was a moment of silent doubt among the 12 wriggling boys.

"That's what I heard, somewhere," came the hesitant reply.

"Well, what if you see lots of trees and not all of them have moss on the same side?"

Silence.

"What about the Sun?" prompted the scoutmaster.

Silence.

"Come on now, you all know where the Sun rises each morning, don't you?

Silence.

Then a weak and unsure response came from the back of the room. "The Sun always rises in the east?"

"And where does it set?"

"In the west … I think," said the young tenderfoot, a first-year scout.

"Right!" said the scoutmaster. "So, if you see the Sun rise or set, you can tell true east or west. That means when we go on our campout this weekend, we'll try to find our way without compasses by using the Sun."

The weekend came. All the boys went out to the campsite, set up their tents and prepared for a hike the next morning. It was a lovely early June morning, not a cloud in the sky, and the Sun came up with a beautiful golden glow on the horizon.

"There it is boys," said the scout leader, "And what direction is that?"

"East?" the boys responded in unison.

"Yep! And now let's just check it out on the compass, so you know that it's true."

Compasses came out of pockets and backpacks and there was a lot of moving around and pointing of the little instruments in all directions. There was also a lot of mumbling.

"Whoops," said one of the boys. "It looks like it is quite a bit north of east to me."

"Let's see how you are reading that compass," said the leader who brought his own compass over next to him and looked in the direction of the rising Sun. Then, he frowned. "Must be something wrong with our compasses."

But, no, each and every scout's compass reported the same result. The Sun was definitely coming up to the north of east, not directly east.

"Well, at least we know that tonight the Sun will set in the west. Even all of the campfire songs tell us that."

That night as the Sun sank beneath the horizon, the boys, a little less sure of their scoutmaster's prediction, checked it out with their compasses just to make sure. What they found gave them a new mystery to solve, one that would take some time and a lot of observations.

PURPOSE

Astronomical rules are not always correct, especially when they use the word "always." Students should learn that unless they live on or very near the equator, there are only two days in the year that the Sun rises directly in the east—the spring and autumn equinoxes. Another purpose of the story is to help understand the importance of latitude in seasonal solar measurements.

RELATED CONCEPTS

- Equinox
- Solstice
- Latitude
- Earth tilt

DON'T BE SURPRISED

Your students probably have heard that the Sun always rises in the east and sets in the west. Unless they have had the opportunity of mapping the sunrises and sunsets in their area, they are probably unaware that the tilt of the Earth affects the apparent direction of these. They may not know that people in the higher latitudes in the northern hemisphere and the lower latitudes in the southern hemisphere see a different path of the Sun than people at the equator. Finally, your students might not be aware that there is a difference between magnetic north and "true" north.

CONTENT BACKGROUND

It is the tilt of the Earth that causes many of the puzzling patterns that we see of our celestial partners, the Sun and the Moon, as they appear to move about our sky. The Earth does not revolve around the Sun straight up and down but is tilted 23½° from the vertical and remains in the same plane as it makes its revolution around the sun each 365+ days, our solar year. (To visualize Earth's tilt, think of the standing globes where you see the Earth leaning a bit to the left.) This causes the seasons, although it seems that much of the population is still under the impression that we are closer to the Sun in summer than winter, and that this is the cause of the seasons. If you would like to see more about this misconception, view the "Private Universe" video on the Annenberg Channel on your computer: *www. learner.org/resources/series28.html*. This fascinating video will allow you to view how in 1987 Harvard graduates—and faculty—had trouble explaining the cause of the changing seasons (Schneps 1987).

We in the northern hemisphere are actually closer to the Sun in the winter; however, the tilt of the Earth's axis allows more *direct* sunlight to the southern hemisphere, which is in the direct line of the Sun's ray. The direct rays of the Sun

provide more radiation than indirect rays do, so those areas that receive the more direct rays experience warmer weather. Thus, when the people in the northern United States and Canada have winter, people in Argentina have summer.

But this story is about sunrise and sunset and what we expect of these phenomena with which we are so familiar. Our scoutmaster, and so many others of us, failed to notice the changes that occur every day in those celestial paths. A true observer is one who sees things even though he or she is not seeking to see something in particular. Most of us walk right by many natural phenomena every day and miss them entirely. Our senses are focused on whatever we are doing. Things can happen right before our eyes and we miss them because we are not accustomed to being true observers. One example of this is the Sun and Moon as they make their daily paths across our sky.

If the axis of the Earth were not tilted, every day of the year would be an equinox, with every day having an equal number of lighted and unlighted hours. If we parse the word *equinox,* we find it means "equal nights." Day and night—the number of hours lit and unlit by the Sun—is caused by the *rotation* of the Earth on its axis. But since this axis is tilted in the yearly revolution around the Sun, in the northern hemisphere the vernal equinox marks the beginning of the spring season. The number of daylight hours will begin to increase and nighttime hours will decrease. The land and water in the northern hemisphere begins to point more and more toward the direct rays of the Sun, becoming warmer. The Sun seems to rise higher in the sky as it goes through its daily "path" and the point on the horizon that marks its "sunrise" moves further north. It will appear to be above the horizon for longer amounts of time as we approach the summer solstice.

Around June 21 in the northern hemisphere, when the axis of the Earth is pointed more directly at the Sun and it is above the horizon for the maximum number of hours for the year, we celebrate the summer solstice, or the first day of summer. In earlier cultures, this day was called *midsummer*, a time of supposedly magical importance. Ancient cultures marked this day, along with those of the equinoxes and the winter solstice, with monuments such as Stonehenge and its counterparts in the Inca empire of south and central America, as well as in the southwest at Chaco Canyon, in what is now New Mexico. Stones were set to mark the passage of the Sun's rays at sunrise so that the rays passed onto or through special passages between stones on particular days. It seems as though the ancient people were much more aware of the seasonal changes than are we today. Shakespeare wrote a play about this day, called "A Midsummer Night's Dream."

It may be advantageous to imagine that you are watching this celestial pageant from some distant place in space. From your vantage point, you can witness the path of the Earth, with its axis always pointing in one direction, as it revolves around the Sun. You can see the two equinoxes when the Sun's rays fall on the Earth so that exactly half of the Earth is lighted. You would notice that at those points, the Sun would rise everywhere on the globe directly in the east. And you would witness the solstices when either the northern or southern hemispheres are favored by more direct light. You would also notice that in the northern hemi-

sphere, the summer solstice would cause the sunrise to appear to be north of east and that the Sun would appear to make a prolonged path across the daily southern sky. At the opposite end of the orbit in the southern hemisphere, you would notice that the sunrise would appear to be south of east and make a prolonged path through the northern sky. If you still have trouble envisioning this, animation at *www.mathisfun.com/earth-orbit.html.*

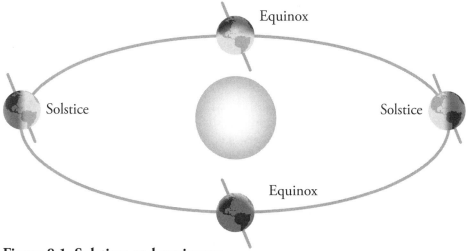

Figure 8.1. Solstices and equinoxes

Our scouts were able to see differences between true east and the northeast sunrise spots, especially in the northern latitudes of Idaho. If they had been further north and closer to the Arctic Circle, the difference would have been even greater. Had they been above the Arctic Circle, the Sun would have never set since the axis of the Earth was pointing directly at the Sun as it made a 24-hour circle around the low horizon. By the way, the scoutmaster might have taken the opportunity to explain that he was using a generality and admit his mistake, but we all know how hard that is!

I want to mention the difference between magnetic north as shown on a compass and "true north," which is the direction from your present location to the pole as indicated on a map. In the southern hemisphere, the difference would be between magnetic south and "true south" of the South Pole. The magnetic poles in this giant magnet we call Earth have shifted as much as 1,200 miles over time. Therefore, the compass will not give you the map-related north or south you need. This is why you need to get a topographic map of the area you are interested in from your local sporting goods store or from the U.S. geological survey. They can be found online at *www.usgs.gov.* Topographic maps give not only contour lines but also a reading of what is called *magnetic declination.* Declination is a number of degrees east or west by which you can make corrections between what is shown on your magnetic compass and the true "map" directions. You either add or subtract numbers from the compass reading to head you in the right map direction. For example, you may find the number -10 or W 10 at your location which means that magnetic north is 10 degrees to the west and you need to *add* 10 degrees to your compass reading to find true north. If the reading is +10 degrees or E 10 degrees, the opposite correction must be made to set you on your correct map direction.

In truth, the boys would have had to refer to their maps before using their compasses to find true east. But the point is still that the sunrise would have been toward the northeast and would probably have been very obvious despite the magnetic declination.

Now comes the kicker. I wonder how many of you realize that the Earth's so-called North Pole is really the *south* magnet pole. The pole is named the North Pole merely because it is in the northern hemisphere. The compass needle is a north pole of a magnet and therefore would be repelled by another north pole. It seeks the north because the north magnetic pole is really the south pole of the Earth's magnetic field. However, it makes no difference in how we view our directions. We still head toward the north pole if we are going north and vice versa. And we still say that the Sun in the northern hemisphere rises each day more toward the northeast as the days progress from the equinoxes to the summer solstice. Don't forget those folks who live in the southern hemisphere who are "standing on their heads" and seeing everything 180° differently than we are.

related ideas from the national science education standards (nrc 1996)

K–4 Changes in the Earth and Sky
- Objects in the sky have patterns of movement. The observable shape of the moon changes from day to day in a cycle that lasts about a month.

5–8 Earth in the Solar System
- Most objects in the solar system are in regular and predictable motion. Those motions explain such phenomena as the day, the year, phases of the moon and eclipses.

related ideas from benchmarks for science literacy (aaas)

K–2: The Earth
- The Moon looks a little different every day, but looks the same again about every four weeks.

6–8 The Earth

- The Moon's orbit around the Earth once in about 28 days changes what part of the Moon is lighted by the Sun and how much of that part can be seen from the Earth—the phases of the Moon.

USING THE STORIES WITH GRADES K–4

The global implications of this story may be too complex for very young children. However, it should raise some interesting questions about the path of the Sun for even the youngest children and give you an opportunity to do some scientific observation and recording. If you read the introduction in this book you will see how a second-grade teacher used another story, "Where are the Acorns?" to guide her students toward understanding the pattern of the Sun's shadows over a whole school year. Some of the techniques used by the teacher could be applied to this story since many of the concepts involved are similar in both. If you are able to obtain a copy of *Everyday Science Mysteries* (Konicek-Moran 2008), the chapter "Where Are the Acorns?" has a great deal of information about using the story with younger children. I would also suggest that you purchase a copy of *The Old Farmer's Almanac* (Yankee Publishing) for the current year. It is very inexpensive and has tables of astronomical events that can be adjusted for your specific location. It contains accurate tables for sunrise, sunset, moonrise, moonset, solar and lunar eclipses, length of day, and many other useful data for teaching any astronomical topic.

With young children, the first step toward understanding that the Sun travels in a predictable pattern lies in observing shadows and learning about how they can change direction and shape. Have children place a pencil in a vertical position on their desks using a piece of clay to hold the pencil in place. Give each team of children a flashlight and try to dim the lights in the classroom. Let them experiment with the flashlight and the shadows created by the pencil when they move the light source. Ask them to do the following:

- Make the shadow move toward your left hand.
- Make the shadow move toward your right hand.
- Make the shadow get longer.
- Make the shadow get shorter and shorter until it disappears.

Have students record or draw where their light source was in each case. Then ask them to help you create a list of things they have learned about shadows. Post these on chart paper and if there are disagreements, test them to settle the discrepancies.

At this point, they are usually ready to go outside and observe their own shadows at different times of the day. They can also play "shadow tag" where they must tag another player's shadow with their own. All of these activities will make them more familiar with how shadows are formed and how they change.

Moving to shadows made by the Sun is the next step. They should now place a stick on level ground so that it is vertical. This is known as a *gnomon*. The word comes from the ancient Greek and means indicator. It is the part of the sundial that casts the shadow. Once they see that the Sun is their light source, the motion and length of the shadows will correspond to their activities with the flashlight. The big difference is that they cannot control the Sun like they did the flashlight, so they must wait for the Sun to move and change the shadow. This means that they must record the shadow position and length on a paper. The easiest way to do this is to stick the gnomon through the middle of a sheet of paper and secure the edges of the paper with rocks or toothpicks. At various intervals during the day, children can record the time and draw over the shadow to record their lengths and directions.

As children get used to the apparatus, and perhaps for children who are a bit older, they can see that in the northern hemisphere, after the equinox in March, the morning shadows will retreat a bit more toward the south each day in comparison to where they were earlier in the month at the same time. This can be translated to mean that the Sun is further north in the sky so that the opposing shadow appears further south on the paper. Of course, in the southern hemisphere the directions would be reversed.

Figure 8.2. Gnomon

They may also discover that as the summer months approach, the shortest shadow of the day becomes shorter meaning that the Sun is becoming higher in the sky at midday. If you should happen to teach on Key West in southern Florida, you are only one degree north of the Tropic of Cancer which is 23.5° north of the equator. On June 21(the summer solstice) the Sun will be very close to being directly overhead at midday on that date. But that is as far north as the overhead Sun will get. In other words, the myth that the Sun is directly overhead everywhere at noon is false.

Reasons for the seasons should probably be reserved for the middle school years and we will continue suggestions for this in the next section.

USING THE STORIES WITH GRADES 5–8

My experience with students in grades 5–8 is that some, but not all, are able to use the spatial relationships necessary to understand such things as the reasons for the seasons and the phases of the moon. I do believe that it is worth a try but do not be surprised if they finally "get it" sometime later in someone else's class. You can be satisfied that you did your best and that whatever you did laid the groundwork for the final "Aha!" that is bound to come later.

I believe it would be helpful to you to view the "Private Universe" video (Schneps 1987) mentioned previously and see it for yourself before you begin to work with your students. You will see students, not much older than your own, struggle with the ideas of the interactions of the sun, moon, and earth. You'll notice that one of the students quickly used props to help her understand how the three celestial globes related to one another. Although not all students have

a tactile learning style, it usually helps most students to play with objects and to maneuver them to help them to build a visual model of the positions that cause things like the seasons, moon phases, and eclipses. Even though this is not the main purpose of this story, it helps if the students have some understanding of how the Sun's rays, as they impact the Earth, change over a period of a year.

If you would like to introduce your students to orienteering or using maps to find their way, you will need to teach them compass skills and make them aware of the magnetic declination issue described above. Sometimes, orienteering is a great way to engage students in discovering the intricacies of the magnet upon which we spend our lives.

Students in middle school should now be able to finish the story about the scout trip and make sense of the problem caused by the discrepancy between the scoutmaster's general statement and what they found with their compasses. They may want to do some checking on the equinox to see if it is true as well that the sun rises directly in the east. After all, checking patterns for oneself is a major part of science. Also point out the importance that it is a June camping trip in the story (well past the solstice).

related NSTa Press BOOKS and NSTa JOURNal articles

Keeley, P, 2005. *Science curriculum topic study: Bridging the gap between standards and practice.* Thousand Oaks, CA: Corwin Press.

Keeley, P., F. Eberle, and L. Farrin. 2005. *Uncovering student ideas in science, volume 1: 25 formative assessment probes.* Arlington, VA: NSTA Press.

Keeley, P., F. Eberle, and J. Tugel. 2007. *Uncovering student ideas in science, volume 2: 25 more formative assessment probes.* Arlington, VA: NSTA Press.

Keeley, P., F. Eberle, and C. Dorsey. 2008. *Uncovering student ideas in science, volume 3: Another 25 formative assessment probes.* Arlington, VA: NSTA Press.

Keeley, P., and J. Tugel. 2009. *Uncovering student ideas in science, volume 4: 25 new formative assessment probes.* Arlington, VA: NSTA Press.

Konicek-Moran, R. 2009. *More everyday science mysteries: Stories for inquiry-based science teaching.* Arlington, VA: NSTA Press.

Konicek-Moran, R. 2010. *Even More Everyday Science Mysteries: Stories for inquiry-based science teaching.* Arlington, VA: NSTA Press.

references

American Association for the Advancement of Science (AAAS).1993. *Benchmarks for science literacy.* New York: Oxford University Press.

Driver, R., A. Squires, P. Rushworth, and V. Wood-Robinson. 1994. *Making sense of secondary science: Research into children's ideas.* London and New York: Routledge Falmer.

Gilbert, S. W., and S. W. Ireton. 2003. *Understanding models in earth and space science*. Arlington, VA: NSTA Press.

Keeley, P., and J. Tugel. 2009. *Uncovering student ideas in science, volume 4: 25 new formative assessment probes*. Arlington, VA: NSTA Press.

Konicek-Moran, R. 2008. Where are the acorns? In *Everyday science mysteries: Stories for inquiry-based science teaching*, 39–50. Arlington, VA: NSTA Press.

Konicek-Moran, R. 2009. *More everyday science mysteries: Stories for inquiry-based science teaching*. Arlington, VA: NSTA Press.

National Research Council (NRC). 1996. *National science education standards*. Washington, DC: National Academy Press.

Oates-Brockenstedt, C., and M. Oates. 2008. Earth science success: 50 lesson plans for grades 6–9. Arlington, VA: NSTA Press.

Schneps, M. 1986. *The Private Universe Project*. Harvard Smithsonian Center for Astrophysics.

United States Geologicical Survey. *www.usgs.gov*

Yankee Publishing. *The old farmer's almanac*, published yearly since 1792. Dublin NH: Yankee Publishing.

BIOLOGICAL SCIENCES

Core Concepts	Lookin' at Lichens	Baking Bread	Springtime in the Greenhouse	Reaction Time	Seedlings In A Jar
Fungi	X	X			
Algae	X				
Symbiosis	X				
Spores	X	X			
Reproduction	X	X	X		X
Life cycles	X	X	X		X
Yeast		X			
Metabolism	X	X	X		X
Chemical Change	X	X	X		X
Physical Change	X	X	X		X
Nutrition	X	X	X		X
Germination			X		X
Photosynthesis	X		X		X
Nervous System				X	
Reaction Time				X	
Stimuli				X	
Responses				X	
Averages					
Systems	X	X	X	X	X
Open Systems	X	X	X		
Closed Systems					X
Atmosphere	X	X	X		X
Experimental Design	X	X	X	X	X

CHAPTER 9
LOOKIN' AT LICHENS

Rick and Jeannie were walking through the woods one morning. Suddenly, Jeannie stopped in her tracks and said, "Rick, look at this tree! It has white bark. I don't know of any trees that have completely snow-white bark, except maybe a birch. But this isn't a birch!"

Rick took a closer look and exclaimed, "Wow! You're right. This is no ordinary tree."

Now, Jeannie and Rick never took a walk in the woods without their magnifying glass and so they decided to look more closely at the strange tree with the white bark.

"It looks like it is one big white scab," said Rick, "and it completely covers the trunk of the tree."

"If you look at it closely, you can see that there are little bumps on it and some little black lines that look like they are raised above the surface!" Jeannie added, peering through the glass. "I wonder if we can peel it off and take some home for a better look under strong lights?"

"Let's try," said Rick. "Oh, man, this stuff is stuck right to the bark of the tree and won't come off unless we take some bark too."

"Well, I don't think we will hurt the tree if we take just a little bark," said Jeannie. And so they did.

But the walk was not over and they began to see all sorts of interesting things growing on the trees. But they were not all alike! Some looked like little flowers but were green all over. Some looked like plants that had overlapping scales. Many resembled the white stuff they had spotted on the first tree, but were red, pink or yellow. Once they began to notice them, it seemed like they were everywhere. They were on leaves, rocks, and even on the ground! They took a lot of samples and found that some of the little green things came off the trees where they were growing without much, if any bark. The ones on the rocks would not come off easily at all and the same was true of the ones on the leaves. When they got home, they found some on the door of the woodshed. Now they began to see them almost everywhere.

"Why haven't we noticed them before, I wonder?" asked Rick. "Now that we have noticed them, it seems like we can't find anything that doesn't have some of them growing on it. Look, there is even some on the railing of this stair up to the house. I wonder what they are?"

Jeannie and Rick knew Rebecca, a biologist at the local nature center and went to ask her. She looked at their samples and immediately said, "Those are lichens."

"Like-ums?" said Rick, "That's a funny name, but we do like 'em."

"No, lichens," said Rebecca. "L-I-C-H-E-N-S. They are very special kinds of organisms that are made up of two different kinds of living things. They live together, dependent on each other, in a way."

"How can we find out more about them?" the two young scientists asked together.

"Well, I can help you some, but I think you can learn a lot just by looking at them under a microscope or a magnifying glass. You can keep on collecting them. Maybe drawing them will help too."

PURPOSE

Lichens are everywhere, yet most people fail to notice them because they are so familiar. This story was written to help persuade teachers to acquaint their students with these unique forms of life. Many biology teachers, including myself, tend to gloss over the study of lichens and many of the other simple plants, even though lichens are universally available in virtually every environment, including urban centers. I hope that this story will help more students appreciate and become interested in them.

related CONCEPTS

- Fungi
- Algae
- Symbiosis
- Spores
- Reproduction
- Life cycles

DON'T BE SURPRISED

Don't be surprised if your students have never noticed or expressed interest in the lichens. Lichens are not usually flashy, although some of them have beautiful patterns and colors. Your students are probably not aware of the kind of relationship the fungi and the algae have. Older students will probably have some knowledge of plants or animals living together in some form of mutually dependent relationship (*symbiosis*) such as the bacteria and protozoa in the guts of termites. But the association between the fungi and algae in the lichens is an entirely unique relationship, well worth studying.

CONTENT BacKGROUND

A *lichen* is a composite of a fungus and another organism that is capable of producing food through photosynthesis. The latter may be green algae or *cyanobacteria* (blue-green algae) or sometimes both. When two organisms have a biological relationship it is called *symbiosis,* and the partners are called *symbionts.* In lichens, the fungal part is called the *mycobiont,* and the algae the *phytobiont* or sometimes a *photobiont.* Lichens are named after the fungus partner since observing and classifying the algae is not practical because they are hidden within the *thallus,* or the body of the fungus. It was not until 1867 that anyone even thought that the lichen might be symbiotic, because the idea of two organisms living together as such was unheard of. It took until 1939 before the true nature of the lichen was proved and then accepted by the scientific community. You might find *Lichens of North America* (Brodo, Sharnoff, and Sharnoff 2001) of interest.

Neither fungi nor the two types of algae are members of the plant kingdom, but each belong to their own: fungi in the kingdom Fungi, the cyanobacteria (blue-green algae) in the kingdom Monera, and the green algae in the kingdom Protista. Monerans are single-celled organisms that have no membrane around their nucleus. Protista is a kingdom that seems to encompass everything that doesn't fit anywhere else, ranging from tiny protozoa to 30-meter-long kelp. It is important to realize that classification is somewhat arbitrary since it can change over time depending upon how the scientific community decides what traits define the organisms.

The algae in lichens provide sugars for the fungi, while the fungi provide protection from ultraviolet (UV) light for the algae and some predators that could kill or damage them. The algae live between two or three layers of fungi, like the filling in a peanut butter and jelly sandwich, which is a good analogy for the lichen form (or *morphology)*. The "bread" on top would correspond to what is called the *upper cortex* of the fungus layer, with closely packed fungal cells acting like skin. Beneath lies "jelly," the fungal filaments called *hyphae* in which algae are embedded. These hyphae may grow small tubes that reach into algae to extract the sugars produced by the algae's photosynthesis. Below that is the "peanut butter," which corresponds to the *medulla,* a fungal layer that is not so densely packed as the outer layers, and where many of the living functions of the lichen occur. Finally, in most (but not all) lichens, is the bottom layer of "bread," another tightly packed protective fungus layer called the *lower cortex.* This layer often contains structures that help the lichen attach to its host. Most fungi will combine with only one kind of algae but the algae are not as strict and may cohabitate with several different kinds of fungi.

Something called *morphogenesis* happens when the lichen and algae "marry." Neither organism is the same as it would be if it were found alone. Also, neither of the two organisms that make up lichen can be found living in isolation in the natural state. Only in the laboratory can the two be separated and examined as individuals.

Diagram of the Lichen Structure

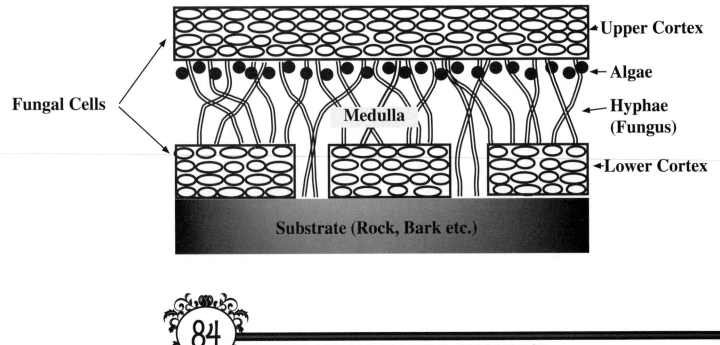

Even here, the resulting organisms are amorphous and fragile. Thus, each lichen is a totally unique organism, named, as stated before, for the fungal partner.

Even though we were taught in school that lichens were the epitome of a mutualistic symbiosis, recent research has shown that few if any lichen species are equal partners in the symbiosis (Brodo, Sharnoff, and Sharnoff 2001). *Mutualism* occurs when two or more organisms interact in a way that is beneficial to all parties. You and your students may immediately think of bees and flowers as examples. Flowers provide nectar for the bees that spread the pollen from flower to flower. Another example of mutualism is the protozoa that live in the gut of termites. The termites offer food and protection while the protozoa help to digest the cellulose (wood products) the termites ingest. The partnership in lichens ranges from mild parasitism by the fungi upon the algae to outright destructive behavior. Some fungi will actually kill their partner algae over time, but usually the photosynthetic partner will reproduce fast enough to keep ahead of the fungal aggressiveness. Lichenologist Trevor Goward described lichens as fungi that have "discovered agriculture" (Brodo, Sharnoff, and Sharnoff 2001, p. 4).

In lichens, the photobiont (alga) can reproduce sexually or asexually (but mostly asexually). The mycobiont (fungi) reproduce by means of spores that can likewise be the result of sexual or asexual activity. When the windborne spores are produced asexually, they often carry some of the photobiont material with them. The spores resulting from sexual activity must find suitable algae with which they can combine in order to become viable. If you look closely, you can see the cup-like *apothecia* or long, dark, narrow ridges called *lirellae* across the surface. Both of these structures contain spores and are often used to identify the species of the

Lichen Spore Structures

lichen. Lichens can also undergo vegetative propagation by having pieces of the thallus break off and become windborne to another location. This way, alga and fungus are kept together.

Lichens come in many forms and colors. *Crustose* (crust-like) lichens look as though they are glued to the bark of trees, leaves or rocks. This is likely the form of lichen found by Rick and Jeannie covering the tree. It is impossible to take them off in their entirety without taking some of the substrate (like the bark). We see only the upper parts of the thallus. The *foliose* (leafy) form of lichen is a little looser in its hold on the substrate, so we can lift and see both the upper and lower layers of the thallus. *Squamulose* (scale-like) forms show many overlapping parts of the thallus. The most dramatic lichen forms, the *fruticose* (shrubby) stand out from the substrate and may even look like moss or vines. One of these fruticose lichens is probably the most well known *and* misnamed, the so-called "reindeer moss," which is really a lichen (*Cladonia spp*).

Lichens are very difficult to identify to species levels. In order to distinguish one from another they often have to be keyed out by the chemicals that they produce.

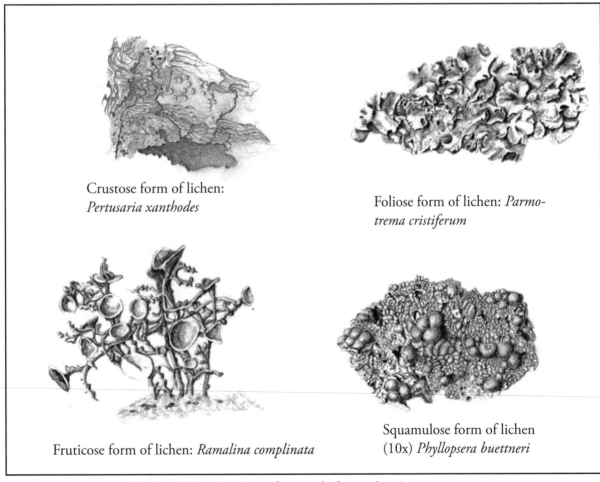

Crustose form of lichen: *Pertusaria xanthodes*

Foliose form of lichen: *Parmotrema cristiferum*

Fruticose form of lichen: *Ramalina complinata*

Squamulose form of lichen (10x) *Phyllopsera buettneri*

(Thanks to Rick and Jean Seavey for original reference photos)

These *metabolites*, as they are called, can be recognized through various chemical tests and through a process called *chromatography* that separates out chemicals on either paper or gel by using a solvent. These chemicals can also glow in the presence of UV light. Each species of lichen has a distinctive color emission when you bathe it in UV light, which is yet another way of identifying the organism.

Lichens know no bounds when it comes to climate or altitude. They range from sea level to mountaintops and from the tropics to the poles. They are found on any type of surface, including tree bark, leaves, plastic, rocks, soil, unwashed vehicles, and even on some insects.

Lichens provide food for many browsing animals and are capable of fixing nitrogen from the atmosphere into compounds useful by other organisms in the environment. They are also a "canary in the mine" in that they absorb pollutants and radioactivity so that their lack of health or even disappearance can be seen as a warning of polluted surroundings.

One interesting characteristic of lichens is that they can endure long periods without water and become revived quickly once water is restored. Therefore they are very resilient to climate and weather changes. If you find dried-up looking lichen, you can douse it with water and within a few minutes see the color return.

Despite the fact that they have been so often ignored, these organisms should invite a great deal of scrutiny because of their diversity and uniqueness, as well as their importance to the ecosystems of the world.

related Ideas From The National Science education Standards (NrC 1996)

K–4: *The Characteristics of Organisms*

- Organisms have basic needs
- Organisms can survive only in environments in which their needs can be met. The world has many different environments, and distinct environments support the life of different types of organisms.

K–4: *Life Cycles of Organisms*

- Many characteristics of an organism are inherited from the parents of the organism, but other characteristics result from an individual's interaction with the environment.

K–4: *Organisms and Their Environment*

- An organism's patterns of behavior are related to the nature of that organism's environment, including the kinds and numbers of other organisms

present, the availability of food and resources, and the physical characteristics of the environment.

- All organisms cause changes in the environment where they live. Some of these changes are detrimental to the organism or other organisms, whereas others are beneficial.

5–8: *Structure and Function in Living Systems*

- Living systems at all levels or organization demonstrate the complementary nature of structure and function. Important levels of organization for structure and function include cells, organs, tissues, organ systems, whole organisms, and ecosystems.
- All organisms are composed of cells—the fundamental unit of life. Most organisms are single cells; other organisms, including humans are multicellular.
- Cells carry on the many functions needed to sustain life. They grow and divide, thereby producing more cells. This requires that they take in nutrients, which they use to provide energy for the work that cells do and to make the materials that a cell or an organism needs.
- Specialized cells perform specialized functions in multicellular organisms. Groups of specialized cells cooperate to form a tissue, such as a muscle. Different tissues are in turn grouped together to form larger functional units, called organs. Each type of cell, tissue, and organ has a distinct structure and set of function that serve the organism as a whole.

5–8: *Reproduction and Heredity*

- Reproduction is a characteristic of all living systems; because no individual organism lives forever, reproduction is essential to the continuation of every species. Some organisms reproduce asexually. Other organisms reproduce sexually.

5–8: *Populations and Ecosystems*

- A population consists of all individuals of a species that occur together at a given place and time. All populations living together and the physical factors with which they interact compose an ecosystem.
- The number of organisms an ecosystem can support depends on the resources available and abiotic factors, such as quantity of light and water, range of temperatures, and soil composition. Given adequate biotic and abiotic resource and no disease or predators, populations (including humans) increase at rapid rates. Lack of resources and other factors, such as predation and climate, limit the growth of populations in specific niches in the ecosystem.

related ideas from benchmarks for science literacy (aaas)

K–2: *The Living Environment*

- Some (organisms) are alike in the way they look and in the things they do, and others are very different from one another.
- (Organisms) have features that help them live in different environments.

K–2: *The Interdependence of Life*

- Living things are found almost everywhere in the world. There are somewhat different kinds in different places.

3–5: *The Living Environment*

- A great variety of kinds of living things can be sorted into groups in many ways using various features to decide which things belong to which group.
- Features used for grouping depend on the purpose of the grouping.

3–5: *The Interdependence of Life*

- Organisms interact with one another in various ways besides providing food.

6–8: *The Living Environment*

- Similarities among organisms are found in internal anatomical features, which can be used to infer the degree of relatedness among organisms. In classifying organisms, biologists consider details of internal and external structures to be more importantt than behavior or general appearance.

6–8: *The Interdependence of Life*

- Two types of organisms may interact with one another in several ways; they may be in a producer/consumer, predator/prey, or parasite/host relationship. Or one organism may scavenge or decompose another. Relationships may be competitive or mutually beneficial. Some species have become so adapted to each other that neither could survive without the other.

USING THE STORIES WITH GRADES K–4

The best way to start this exploration is to go on a field trip to the backyard, schoolyard, or anywhere there are trees, rocks, or objects upon which lichens grow. Take the trip yourself, first. If lichens are not available on school grounds, you can usually find them growing on tombstones in graveyards, particularly ones with limestone grave markers. Often, in older communities and even older schools, lichens readily grow on buildings. If you live in a mountainous region, the rocks will have lichens growing on them in abundance.

Most children and adults will have seen lichens, but think that they are stains on the rocks and trees. Closer observation will show that many of the "stains" are really living lichens. Each child should have a magnifying glass and a notebook to draw them in. Think about having a scavenger hunt, with teams of kids sent out to find and record the most lichens. For those lichens that can be taken back to the classroom, a digital projecting microscope can reveal the unique qualities of each. There are now digital microscopes on the market that allow you to save pictures to a computer. If your school has online capabilities, you can find many pictures of lichens on the internet through your search engine.

Another activity involves the careful scraping of the top cortex from the lichen to reveal the green algae layer below. This can be accomplished only with careful dexterity and should probably be done by the teacher and shown to the students in the best way you have at your disposal (microscope, digital microscope, or hand lens).

While looking for the lichens in the local environment your students can probably think of some interesting questions to investigate:

- What is the most common color of lichens we have found?
- What kinds of shapes did the lichens exhibit?
- What kinds of trees do lichens seem to grow on most often?
- What kinds of trees do lichens seem to grow on least often?
- How many different colors did we find?
- Do they grow on any particular side (direction) of trees?
- Are there different kinds of lichens on the same tree or rock?
- What was the most common kind of lichen found?
- If animals were in or on the lichens, what kind were they?
- Within a set area, how many lichens were found and what kinds?

As you can see, there are many questions about the location, type, and physical attributes of lichens. Looking at lichens that are brought back to the classroom can provide some interesting observations too. Be careful, lichens grow slowly and can be damaged, so only bring home as many as you absolutely need. Drawings of the lichens plus information gathered on the field trip should be carefully recorded in their science notebooks.

USING THE STORIES WITH GRADES 5–8

A field trip to the immediate neighborhood is in order for the middle school group but you may wish to organize the trip differently than for younger students. Middle school students can set up an area where the lichens are most prevalent and then identify the trees where the most lichens are found. They might answer some of the same questions listed above and also add such questions as:

- What kinds of lichens are most prevalent on what types of bark?
- What types of bark are lacking lichens?
- Are colored lichens or plain lichens most often found on soil or on the rocks?
- Are there any lichens that seem to be trying to occupy the same space?
- What kind of lichens (crustose, foliose, squamulous, or fruticose) did they find?

If microscopes are available, have the students look at the lichens under low power and find the apothecia and lirallae if they are present. If the lichen is dry, have them place a few drops of water on the lichen thallus and then watch what happens as the water soaks in. Under microscopic view, they can scrape away a bit of the top of the thallus and get a view of the green layer containing the algae or cyanobacteria. They can also look for small organisms that might be lurking in the tangle of the lichen structure.

This may be an eye opening experience for most students and possibly for you as a teacher. Lichens are usually glossed over by most biology teachers, so this may be one of the few opportunities for your students to look at them closely. I can almost guarantee that once they are seen, they will hold your class's interest.

related NSTa Press Books and NSTa Journal articles

Keeley, P, 2005. *Science curriculum topic study: Bridging the gap between standards and practice.* Thousand Oaks, CA: Corwin Press.

Keeley, P., F. Eberle, and L. Farrin. 2005. *Uncovering student ideas in science, volume 1: 25 formative assessment probes.* Arlington, VA: NSTA Press.

Keeley, P., F. Eberle, and J. Tugel. 2007. *Uncovering student ideas in science, volume 2: 25 more formative assessment probes.* Arlington, VA: NSTA Press.

Keeley, P., F. Eberle, and C. Dorsey. 2008. *Uncovering student ideas in science, volume 3: Another 25 formative assessment probes.* Arlington, VA: NSTA Press.

Keeley, P., and J. Tugel. 2009. *Uncovering student ideas in science, volume 4: 25 new formative assessment probes.* Arlington, VA: NSTA Press.

Konicek-Moran, R. 2008. *Everyday Science Mysteries: Stories for inquiry-based science teaching.* Arlington, VA: NSTA Press.

Konicek-Moran, R. 2009. *More everyday science mysteries: Stories for inquiry-based science teaching.* Arlington, VA: NSTA Press.

Konicek-Moran, R. 2010. *Even More Everyday Science Mysteries: Stories for inquiry-based science teaching.* Arlington, VA: NSTA Press.

references

Brodo, I. M., S. D. Sharnoff, and S. Sharnoff. 2001. *Lichens of North America.* New Haven: Yale University Press.

Driver, R., A. Squires, P. Rushworth, and V. Wood-Robinson. 1994. *Making sense of secondary science: Research into children's ideas.* London and New York: Routledge Falmer.

CHAPTER 10
BAKING BREAD

There aren't a lot of things that smell better than fresh bread baking in the oven. It makes your mouth water and you dream of putting some fresh butter or jam on the warm bread, then enjoying a wonderful snack. MMMMM! Fresh bread is crusty on the outside, soft and tasty on the inside. At least that is what Rosa and Sofia were thinking as they looked at the new bread machine at Grandma's house where they were visiting one day after school. Grandma's house usually smelled great every day because she was a baker of cookies, pies, and all sorts of fantastic things to eat. Today though, the adventure was to try out the new bread machine that she had received as a gift. It not only mixed the ingredients and kneaded the bread but it baked it too, all without any person touching the dough! All you had to do was to pour in the ingredients, then sit back and wait for the tasty treat to come out of the machine.

Grandma used her usual recipe, which consisted of yeast, sugar, flour, salt, lukewarm water, and milk. After she had poured in the yeast, they closed the machine and sat back to let it do its work. Since it took a couple of hours, they played three games of scrabble. Then, when they heard the beep of the machine and smelled the delicious odor of the bread, they ran to the kitchen to check it out. Grandma pulled out the bread and laid it on the cutting board. But, wait! It didn't look like bread at all. It resembled a brown brick! And it felt like one, too, when it had cooled enough to lift it. This was not what they had been waiting for and that's for sure.

"That machine is not very good if this is what it turns out," said Sophia.

"It's a big disappointment if you ask me," said Grandma. "I wonder what went wrong. I followed the directions every step of the way. I'll tell you what—tomorrow, I'll try again. You can come over to help me sample it. It has to work, or else it goes back!"

The next day, they all watched as Grandma followed the directions and used the same ingredients from the cupboard she had used the day before. And lo and behold, the same disappointing brick emerged from the machine.

"Okay, that's it, the machine goes back," said Grandma, "unless something else is wrong. Hmmm. I wonder if there could be some problem with the ingredients."

"What could be wrong with flour and water and milk and yeast?" asked Rosa.

Grandma held her chin in the palm of her hand and said, "Well, the problem is that the bread did not rise. I think I know which ingredient might be the scoundrel! And we can find out which one, very easily."

So, Grandma checked the expiration dates on all of the ingredients and they used these data to design the experiment.

The next day, the fresh strawberry jam tasted great on the light and fluffy bread from the machine.

PURPOSE

This story is meant to show students the importance of leavening agents in making baked goods. Yeast, a living organism (a fungus), is necessary for baking risen yeast breads. This fungus can be ineffective if it is not healthy. The story will also give directions on proofing yeast so that the sad outcome described in the story is less likely to happen.

RELATED CONCEPTS

- Fungi
- Yeast
- Metabolism of living things
- Chemical change and physical change

DON'T BE SURPRISED

Your students may not be aware of the importance of leavening agents in the preparation of baked goods. They may be familiar with the substance called *yeast* but not realize that it is a living organism and produces materials that are essential for the rising of baked goods by reacting with the sugars in the recipes to form carbon dioxide gas and alcohol. Being a living organism means that it has the needs of any living thing such as air, water, and warmth. Like all living organisms, it has a life cycle and therefore can die.

CONTENT BACKGROUND

Baking bread is one of the most universal forms of food preparation. It dates back to the earliest recorded history and probably existed before that. We have stories from the Old Testament about the Israelites being liberated from bondage and being so rushed that they were unable to take the time to allow leavening to take place in their bread. To commemorate that event, Jewish people today eat unleavened bread during the Passover holiday.

Bread making probably goes back as far as the Stone Age where wild grass seeds were ground and mixed with liquids and baked on flat stones. Modern wheat, as we know it, has no direct wild antecedent and was probably created by selective breeding of several wild ancestors. It is hard to tell exactly when leavening was discovered to create bread in the shapes we find common today. But we do know that the Egyptians used leavening as long as 5,000 years ago. I confess to pondering a good deal about how these discoveries were made in the early years of civilization. For instance, who thought of eating the first raw oyster? I suspect that as people began to use wild animals and plants as food, some mistakes were made and some brave souls became very ill or even died from their toxins which provided valuable data for those who were not such brave souls. The history of food and agricultural production is fascinating, to say the least.

Leavening of baked products always involves the production of a gas, usually carbon dioxide. The bubbles of this gas expand during baking and cause what we call the "rising" of the bread dough. *Yeast*, a single-celled fungus called *Saccharomyces cerevisiae*, feeds on starches, which it first breaks down into simple sugars. As it digests these sugars, it produces carbon dioxide and alcohol. In the place of yeast in certain baking recipes, several other chemicals can be substituted. *Baking soda* is a naturally occurring chemical (sodium bicarbonate) that forms carbon dioxide when it comes in contact with a wet acid. So any recipe using baking soda must have some acid, like buttermilk or sour milk, in order for the carbon dioxide to form. *Baking powder* is a combination of baking soda and acid salts, so no other acidic ingredients are needed. *Sourdough starter* is a bacterium, the *lactobacillus* culture, mixed with yeast, which is more effective with the heavier rye flours. The bacteria feed off the products produced by the yeast, releasing carbon dioxide as well.

You are probably familiar with the mixing of vinegar and baking soda to produce a bubbling chemical reaction, the perennially favorite "volcano" scenario of science fairs throughout history. Sometimes a chemical formula helps us to see the chain of events that occurs in a reaction. The chemical formula for this is as follows:

$$CH_3COOH \text{ (vinegar)} + NaHCO_3 \text{ (sodium bicarbonate-baking soda)} \rightarrow CH_3COONa \text{ (sodium acetate)} + H_2CO_3 \text{ (carbonic acid)}$$

That last product, carbonic acid, quickly decomposes into carbon dioxide and water:

$$H_2CO_3 \rightarrow H_2O + CO_2 \text{ (becoming bubbles, which you see)}.$$

Try mixing these two chemicals in a soda bottle, then swish them around and quickly place a deflated balloon on the bottle opening. The carbon dioxide will soon inflate the balloon. You can test for carbon dioxide by "pouring" the gas from the balloon onto a candle flame. The carbon dioxide will quickly put out the flame.

When yeast encounters sugar in a warm environment, it becomes very hungry and immediately begins to *ferment* the sugar into alcohol and carbon dioxide. The balanced equation for this fermentation is:

$$C_6H_{12}O_6 \text{ (glucose)} \rightarrow 2 \, C_2H_5OH \text{ (ethyl alcohol)} + 2CO_2$$

In other words, glucose is decomposed by the yeast into ethanol (ethyl alcohol) and carbon dioxide. The alcohol acts as a flavoring for the bread and the carbon dioxide helps to plump up the bread dough and it is said to "rise." It is interesting to note that not all sugars react with yeast in the same way. Sucrose reacts best, fructose reacts a little bit, and lactose reacts hardly at all.

The same process happens in fermenting alcoholic beverages, interacting with the sugars in the fruits or grains that are used. An interesting aside is that the historic figure John Chapman (known as Johnny Appleseed) knew about this process

and planted orchards to lease to the pioneers as they traveled westward. The apples that grew on the trees he planted didn't always produce apples that were good to eat. But, he knew that people in the 18th century did not prize apples for eating but instead, fermented the apple juice into alcoholic hard apple cider that was often safer than water to drink and provided them with the relaxation they sought. Even Puritans drank hard cider since there was no biblical admonition against apples, as there was against the fermented grape.

All of these fermentation processes have the hallmarks of a *chemical change* taking place: new compounds are formed and the original components cannot be retrieved by physical means. Each of the original compounds is changed in the process and has no resemblance to its original form. Their physical and chemical properties have changed as well.

It is important to the story that Grandma put in an ingredient that did not react properly in the baking of the bread. Bread that has risen has lots of sponginess, is less dense than flat bread and can be seen to have some spaces in its texture. The most likely suspect is that the yeast, which has a finite shelf life, may not have been viable, and so could not release the carbon dioxide that makes the bread rise. Grandma may also have put the yeast in water that was too hot. Yeast can be killed by temperatures over 40° C. Grandma and the girls can easily test by checking on the temperature of the water and by "proofing the yeast." Proofing yeast is a technique used by most experienced bakers. A small amount of warm water, a little sugar and a package (or the equivalent) of yeast are placed in a container. In five minutes, if the yeast is good, bubbles will form on the surface of the mixture and can then be added to the flour and other ingredients.

The problem of the bread machine "bricks" is solvable and rife with possible hypotheses that can be tested. I will go over some of these methods and hypotheses in the Using the Stories sections.

related ideas from the national science education standards (NRC 1996)

K–4: The Characteristics of Organisms
- Each plant or animal has different structures that serve different functions in growth, survival, and reproduction.

K–4: Properties of Objects and Materials
- Objects have many observable properties, including size, weight, shape, color, temperature, and the ability to react with other substances.

5–8: The Characteristics of Organisms
- Cells carry on the many functions needed to sustain life. They grow and divide thereby producing more cells. This requires that they take in nutrients, which they use to provide energy for the work that cells do and to make the materials that a cell or an organism needs.

5–8: Regulation and Behavior
- All organisms must be able to obtain and use resources, grow, reproduce, and maintain stable internal conditions while living in a constantly changing external environment.

5–8: Properties and Changes in Properties of Matter
- Substances react chemically in characteristic ways with other substances to form new substances (compounds) with different characteristic properties. In chemical reactions, the total mass is conserved.

related ideas from benchmarks for science literacy (aaas 1993)

K–2: Cells
- Most living things need water, food, and air.

K–2: Structure of Matter
- Things can be done to materials to change some of their properties, but not all materials respond the same way to what is done to them.

3–5: Cells

- Some living things consist of a single cell. Like familiar organisms, they need food, water, and air; a way to dispose of waste; and an environment they can live in.

3–5: Structure of Matter

- When a new material is made by combining two or more materials, it has properties that are different from the original materials.

6–8: Cells

- Within cells, many of the basic functions of organisms—such s extracting energy from food and getting rid of waste—are carried out. The way in which cells function is similar in all living organisms.

6–8: Structure of Matter

- Because most elements tend to combine with others, few elements are found in their pure form.
- The temperature and acidity of a solution influences reaction rates. Many substances dissolve in water, which may greatly facilitate reactions between them.

USING THE STORIES WITH GRADES K–4

Cooking and baking appeals to kids of all ages because it is so much a part of their everyday life. Grandma's house always seems to be a magnet; particularly if she is prone to having cookies around and likes to have her grandchildren help with the baking. My grandmother always had a batch of cookies, either in the oven or in the special cookie jar. She didn't have a bread machine and probably would have frowned at the idea of using one. She was a "from scratch" kind of person. Nevertheless, today's children are used to electronic gadgets of all sorts so they might easily be familiar with the bread machine and its idiosyncrasies. But just as some children think that food appears magically in the market, the baking of bread in the home may be a rare event.

I would like to mention at this point the existence of toy ovens known as Easy-Bake Ovens that have been around since the 1960s. They can be used in a classroom with relative safety. Their source of heat is a 100-watt bulb that is contained in a well-insulated box. Small containers of food can be baked in these ovens. The best part is that they are inexpensive and, though they have been around for over 50 years, are still available online or at most stores with toy departments. Your students can try baking bread in the classroom, albeit in small quantities dictated by the size of the oven and pans that it accommodates. Recipes abound on the internet and, I am sure, in the files of your students' parents as well in your own kitchen. However, stores also sell mixes specifically for the Easy-Bake Oven, which

I understand have mixed reviews (pardon the pun!). Other recipes will have to be cut to fit the size of the oven pans but this can be done easily.

The probe "Is It Living?" from Keeley, Eberle, and Farrin (2005) might be informative for you in seeing what, in their environment, your students believe are "living." With this information you can decide how you are going to use the yeast as an example of a living organism.

If you use a yeast recipe, I suggest that you proof the yeast first by adding one tablespoon of white sugar to a half cup of water. The water should be at about 100°º F or 40° C and no warmer or it will kill the yeast. It is best to use a thermometer to be sure. To this water, add the packet of yeast and wait for about five minutes. If the yeast is good, there will be froth or creamy foam on the surface. If not, throw it away and try another packet (and check the expiration date!). The yeast mixture then can be incorporated into your recipe in the appropriate amount and give a good rise to the bread.

If you have a demonstration microscope with 400× magnification, you may be able to show your students the living yeast cells actually going about the business of digesting the sugar and producing gas bubbles. They may also be able to see them budding, or reproducing asexually. It is a remarkable sight and will amaze even the youngest student. Remember, yeast often comes in a freeze-dried form and has to be rehydrated to restore it to its fully living form.

If your students are old enough, you may want to try other leavening agents as called for in recipes. You can demonstrate the mixture of vinegar and baking soda to show them the release of the carbon dioxide. You may want to try the activity mentioned earlier of placing a balloon over the top of a soda bottle with the soda/vinegar mixture. This will also show your students that the carbon dioxide takes up space and has mass. Your students will probably have many suggestions about trying recipes with variations on the amounts and presence of ingredients. Happy baking!

USING THE STORIES WITH GRADES 5–8

If you are not sure if your students have a reasonable understanding of what is living and not, I suggest you give the probe "Is It Living?" by Keeley, Eberle, and Farrin (2005). This information will help you to decide how to proceed with the use of the living yeast and help students classify it as a living organism. Remember, yeast often comes in a freeze-dried form and has to be re-hydrated to restore it to its fully living form.

Most of the suggestions listed above in the K–4 section will also apply to middle school students. If they are capable of creating and understanding chemical equations, they might be taught to balance the equations. They can experiment with different amounts of leavening agents in beakers and describe and measure the results.

There is one easy activity that can be used with baking soda and vinegar that helps students understand the physical concept of *conservation of matter*. Ask them if they can design an investigation to see if there is any weight loss or gain when

the leavening agent and the acid are mixed. They can put the baking powder in the balloon, allowing them to put the balloon on the bottle top before the reaction starts, thus eliminating any gas loss. Since this is a closed system, the resulting chemical reaction will ensue and the end weight should be exactly the same as when all of the individual parts of the system were weighed prior to the reaction.

In middle school, microscopes are often present so that the students can study the yeast reacting to the sugar water. Make a mixture of yeast and sugar in water just like in the proofing test. Once the froth has been on the top of the mixture for about 30 minutes, take a bit off and place it on a slide and, if possible, view it under approximately 400× magnification. Students should be able to see the yeast cells budding, that is, little buds forming on the parent cell. It is sometimes possible to see gas bubbles being formed as they go about digesting and converting the sugar to carbon dioxide and ethanol.

Finally, I recommend the article in the journal *Science Scope* entitled "Bread Making: Biotechnology and Experimental Design" (Sitzman 2003). This article may fill in some of the gaps you may have in helping your students design experiments concerning yeast. The author suggests using what he calls "liquid bread," which is all of the wet ingredients in the bread recipe. This can make things simpler for you in that you can experiment with excluding ingredients, such as sugar when you use yeast to show the need for it in the fermentation process. This is equivalent to proofing yeast except when you leave out the sugar, there is no frothing, and the appearance of frothing is your evidence of leavening. Leaving out the yeast or the sugar will result in a lack of the chemical reaction necessary for the rising of bread dough, which is easily seen in the "liquid bread."

Students may ask about the process called "kneading" which they see in pizza shops and anywhere that bread or bread products are made. The reason for kneading is to distribute the gases in the bread evenly in the dough so that the final product will be as homogeneous as possible. As you can see, the gas produced by leavening agents is a very important part of creating bread or breadlike products from the chemical reactions between the ingredients. You may also challenge your students to identify the process as one created by a chemical, not a physical change. Remember, the hallmark of a chemical change is that the final product cannot be undone and the individual ingredients cannot be recovered.

Everyone will enjoy sampling the product of their investigations if you decide to carry it out to its completion. Happy baking!

RELATED NSTA PRESS BOOKS AND NSTA JOURNAL ARTICLES

Keeley, P. 2005. *Science curriculum topic study: Bridging the gap between standards and practice.* Thousand Oaks, CA: Corwin Press.

Keeley, P., F. Eberle, and J. Tugel. 2007. *Uncovering student ideas in science, volume 2: 25 more formative assessment probes.* Arlington, VA: NSTA Press.

Keeley, P., F. Eberle, and C. Dorsey. 2008. *Uncovering student ideas in science, volume 3: Another 25 formative assessment probes.* Arlington, VA: NSTA Press.

Keeley, P., and J. Tugel. 2009. *Uncovering student ideas in science, volume 4: 25 new formative assessment probes.* Arlington, VA: NSTA Press.

Konicek-Moran, R. 2008. *Everyday Science Mysteries: Stories for inquiry-based science teaching.* Arlington, VA: NSTA Press.

Konicek-Moran, R. 2009. *More everyday science mysteries: Stories for inquiry-based science teaching.* Arlington, VA: NSTA Press.

Konicek-Moran, R. 2010. *Even More Everyday Science Mysteries: Stories for inquiry-based science teaching.* Arlington, VA: NSTA Press.

references

Driver, R., A. Squires, P. Rushworth, and V. Wood-Robinson. 1994. *Making sense of secondary science: Research into children's ideas.* London and New York: Routledge Falmer.

Keeley, P., F. Eberle, and L. Farrin, L. 2005. Is it living? In *Uncovering student ideas in science, volume 1: 25 formative assessment probes,* 123–130. Arlington, VA: NSTA Press.

Sitzman, D. 2003. Bread making: Biotechnology and experimental design. *Science Scope* 36 (5): 27–31.

CHAPTER 11
SPRINGTIME IN THE GREENHOUSE: PLANTING SEASON

It was springtime again. Eddie and Kerry were back hanging out with their mother in the now year-old greenhouse. Mom is a master gardener, and she loved to plant seeds for their vegetable and flower gardens. Even though it was still cool outdoors, the cozy daytime warmth within the greenhouse made it possible for Mom to plant seeds to transplant into their own garden, as well as into her customers' gardens. She could wait and buy ready-to-go plants at the nursery, but that just added to the cost. Growing plants "from scratch" allowed Mom to make a little profit in her gardening business. The greenhouse was unheated except for sunshine during the day, so at night, it cooled off. So Mom had bought a big electric heating pad, which provided nighttime warmth to seed trays placed upon it.

The three of them were standing in the greenhouse with packets of seeds, planting trays, soil, water, and fertilizer.

"Hey," said Mom, "do you want to help me to plant our new crop for the year?"

"Sure," said the kids, who loved watching the little seedlings emerge from the soil. "But what do we need to do to make these seeds sprout?"

Mom liked to help them with vocabulary, so she told them that a more correct word for sprouting was *germinating*. "What do you kids think we should put in the planters along with the seeds so that they germinate and start their new lives strong and healthy?" she asked.

"Well, soil for sure, and fertilizer seems to be important for plants to grow, so I guess fertilizer would be good," said Kerry. "But I'm not sure it is necessary since I heard that the seed has everything it needs to spr...er, germinate right inside."

"I'm pretty sure that we need sunshine to help them germinate, but how does it get down to the seed when it's under the soil?" asked Eddy.

"Good question," said Mom.

"Maybe it is just the sunshine on the top of the soil that does the trick," said Kerry. "But I'm sure they need water."

"Why do you think that?" asked Eddy.

"Well, we did a lot of stuff with seeds at school and we always had to soak them first or they wouldn't open up so we could look at what was inside."

"Okay … that sounds right. But don't they need to be kept warm all of the time?" asked Kerry. "I sorta think so."

"Well, the greenhouse does that during the day and the heating pad will do that at night," said Eddy.

"Hmm. I guess we have some questions to answer here," said Mom. "How do you think we can set up a little experiment to find out what seeds need and don't need to germinate?"

PURPOSE

Children are often under the impression that fertilizer is necessary for the health of plants just as vitamins are for people and animals. However, seeds and plants are entirely different things. It's important to help them realize that most seeds come equipped with all the nutrients they need to germinate and basically only require warmth, oxygen, and moisture to begin growth.

RELATED CONCEPTS

- Nutrition
- Germination
- Life cycles
- Photosynthesis

DON'T BE SURPRISED

Children and adults alike are prone to believe that if a little of something is good, more is better. In the case of this story, your students will probably believe that if fertilizer is good for growing plants, it must be good for germinating seeds. This may come from seeing gardeners plant seeds in fertilized beds, which is a way of making sure that the germinated seeds will have a fertile substrate in which to grow, but has nothing to do with the seeds' ability to germinate. You may be surprised yourself that some seeds need light for germination, a theory that is being verified through new research.

CONTENT BACKGROUND

Seeds are little packages of "ready-to-go plants." They come with a protective coating, food for their first meals and the ability to break open their seed coat and begin life. If they are healthy, they are definitely alive. Many people do not consider them so because they appear dead or unmoving, like rocks, but they have all the characteristics of a living organism. All the while they are waiting to be planted, they are respiring and using the stored food (*endosperm*) until that time that conditions are right for them to germinate. Seeds need at least three things to make this happen: moisture, oxygen and warmth. Taking in moisture (*imbibition*) will help them to soften the seed coat and swell the cells inside the seed, to begin the growth processes directed by the plant's genetic instructions. They need oxygen for *cell respiration*, like all living things. These processes are facilitated within warmer, rather than colder temperature ranges. Until these conditions are met, the seed stays dormant. There are some seeds that also need light in order to germinate, but they are the exception instead of the rule. In some lettuce and barley seeds, light is needed to stimulate the action of gibberellic acid, which sets a complex set of biochemical steps in motion to help the seeds germinate.

Water enters the seed usually through the opening called the *hilum*, where the seed was attached to the wall of the ovary. The process called imbibition swells the seed and breaks down the seed coat, allowing the seed to make connection with its surrounding environment. Water also helps to release the enzymes necessary to break down the stored food (*endosperm*) into usable components of oils, starches, and proteins. This food will help the seed to maintain its life until it has released several parts of the embryonic seed (the primitive forms of root, stem, and leaves) that will eventually make it a free-living, photosynthetic organism.

All this time, the seed needs to carry out *cellular respiration* to consume its stored food supply. This means that the embryo uses oxygen to burn the food and use the by-products. When the seed's protective coat is broken, the seed gains access to this element in the soil.

During germination, the *radical*, or primitive root, emerges, followed shortly by the *hypocotyl,* the primitive stem upon which are found the *cotyledons*, the "seed leaves" that will begin the first photosynthetic processes so that the new seedling can carry on its own metabolism. All of the above parts are present in an embryonic form in the seed before it has germinated. Soon after germination and the establishment of the seedling, the cotyledons will fall off and the new leaves will take over as the plant grows to its fullest potential. For a wonderful little animation of a germinating seed, go to *www.botanical-online.com/animation4.htm*. You may find other animations by searching the internet for "seed germination animation."

After the germination of the seed complete, the new plant will certainly benefit from the fertilizer in the soil, but that was not the question asked in the story. Mom explicitly asked what the seed needs to germinate. I believe that your students will know all there is to know in order to carry out the investigation to see what seeds need or don't need in order to germinate.

related ideas from the National Science education standards (NRC 1996)

K–4: Abilities Necessary to Do Scientific Inquiry
- Ask a question about objects, organisms, and events in the environment.
- Plan and conduct a simple investigation.
- Employ simple equipment and tools to gather data and extend the senses.
- Use data to construct a reasonable explanation.
- Communicate investigations and explanations.

K–4: The Characteristics of Organisms

- Organisms have basic needs. For example animals need air, water and food; plants require air, water, nutrients and light. Organisms can survive only in environments in which their needs can be met.
- The world has many different environments and distinct environments support the life of different types of organisms.
- Each plant or animal has different structures that serve different functions in growth, survival, and reproduction.

K–4: Life Cycles of Organisms

- Plants and animals have life cycles that include being born, developing into adults, reproducing, and eventually dying. The details of this life cycle are different for different organisms.
- Plants and animals closely resemble their parents.

5–8: Abilities Necessary to Do Scientific Inquiry

- Identify questions that can be answered through scientific investigations.
- Design and conduct a scientific investigation.
- Use appropriate tools and techniques to gather, analyze, and interpret data.
- Think critically and logically to make the relationships between evidence and explanations.

5–8: Structure and Function in Living Systems

- Living systems at all levels of organization demonstrate the complementary nature of structure and function. Important levels of organization for structure and function include cells, organs, tissues, organ systems, whole organisms, and ecosystems.

5–8: Reproduction and Heredity

- Reproduction is a characteristic of all living systems because no individual organism lives forever. Reproduction is essential to the continuation of every species. Some organisms reproduce asexually. Other organisms reproduce sexually.

5–8: Diversity and Adaptations of Organisms

- Millions of species of animals, plants, and microorganisms are alive today. Although different species might look dissimilar, the unit among organisms becomes apparent from an analysis of internal structures, the similarity of their chemical processes and the evidence of common ancestry.

related ideas from Benchmarks for science literacy (aaas 1993)

K–2: Scientific Inquiry

- People can often learn about things around them by just observing those things carefully, but sometimes they can learn more by doing something to the things and noting what happens.
- Describing things as accurately as possible is important in science because it enables people to compare observations with those of others.
- When people give different descriptions of the same thing, it is usually a good idea to make some fresh observations instead of just arguing about who is right.

K–2: Diversity of Life

- Some animals and plants are alike in the way they look and in the things they do, and others are very different from one another.
- Plants and animals have features that help them live in different environments.

K–2: The Physical Setting

- Things can be done to materials to change some of their properties, but not all materials respond the same way to what is done to them.

3–5: Diversity of Life

- A great variety of kinds of living things can be sorted into groups in many ways using various features to decide which things belong in which group.
- Features used for grouping depend on the purpose of the grouping.

3–5: Scientific Inquiry

- Results of scientific investigations are seldom exactly the same, but if the differences are large, it is important to try to figure out why. One reason for following directions carefully and for keeping records of one's work is to provide information on what might have caused the differences.
- Scientists do not pay much attention to claims about how something they know about works unless the claims are backed up with evidence that can be confirmed with a logical argument.

6–8: Scientific Inquiry

- If more than one variable changes at the same time in an experiment, the outcome of the experiment may not be clearly attributable to any one of the variables. It may not always be possible to prevent outside variables from influencing the outcome of an investigation but collaboration among investigators can often lead to research designs that are able to deal with such situations.

6–8: Diversity of Life

- Animals and plants have a great variety of body plans and internal structures that contribute to their being able to make or find food and reproduce.
- For sexually reproducing organisms, a species comprises all organisms that can mate with one another to produce fertile offspring.

USING THE STORIES WITH GRADES K–4

You might want to start by giving the probe "Needs of Seeds" in *Uncovering Student Ideas in Science: Volume 2* (Keeley, Eberle, and Tugel 2007). This probe asks students to choose what, among several choices, seeds need to germinate. It will also ask them to explain their thinking in their own words. This type of formative assessment will give you information you can put to use immediately.

Try asking students to list those things they think are absolutely necessary for seeds to germinate. Even with young children, I would suggest using the correct term and making sure they understand what it means, which is where the animations are helpful. Using this list, change each one to a question. Each statement then becomes an investigable question. For instance, children usually say:

- Seeds need sunlight to germinate.
- Seeds need fertilizer to germinate.
- Seeds need heat to germinate.
- Seeds need air to germinate.
- Seeds need water to germinate.
- Seeds need soil to germinate.

Change the first statement to, "Do seeds need sunlight to germinate?" and so on. Keep this list of questions visible during the teaching sequence to allow students to investigate each question and to keep the appropriate statements or delete them when they find out the results.

You might also find useful information in "Seed Bargains" from *Everyday Science Mysteries*, (Konicek-Moran 2008) or the prequel to this story "The New Greenhouse" in *More Everyday Science Mysteries* (Konicek-Moran 2009).

Pick some fast germinating seeds such as zinnia, marigolds, mung beans, or alfalfa. You should get results in a few days with these. If you plant the seeds in the soil close to the edge of a clear plastic cup, you will be able to see the roots mature beneath the surface of the soil. To answer the last question about seeds needing soil, I suggest placing seeds in a plastic sealed envelope or baggie with a moist paper towel. This will show them that soil is not necessary for seeds to germinate. If they haven't grasped the concept of germination versus growing to maturity as a plant, this will help to establish the difference.

Letting the students organize their plantings, control variables, and carry out the investigation will take quite a bit of help from you. They will have to figure out how to manage the seeds in containers so that all other variables are controlled. A good question to ask them if they fail to control variables is, "How will you know whether or not _____ was responsible for the germination?"

Containers without moisture or warmth should not germinate but all others should, and these results will help to decide the important factors necessary for germination. Of course, the seedling in the container with the fertilizer should not come up any faster or healthier than the rest. This should, at a minimum, raise doubts about the necessity of fertilizer in the medium. I say this because some children will hold stubbornly to their prediction and may find some reason to infer that the fertilizer did make a difference. After all, for many people, "Seeing is believing."

This also leads into the critical thinking skill of setting criteria for success. Some will believe that time is the variable that distinguishes success (the fastest germination). Remind them that they are testing to see *if* seeds germinate with or without certain conditions in the medium, not *how fast* they do it. You should strongly emphasize that not all seeds are able to germinate since they are not all equally healthy. This means planting several seeds in each container. The number of seeds that germinate should also not be a determinant of whether or not the experimental container has met all of the needs. The percentage of germinating seeds should remain fairly constant if you use this year's seeds and plant them properly. For a look at suggestions for using trade books for this topic, NSTA members can access "Teaching Through Trade Books: What Happens to Seeds?" (Ashbrook 2005) or you can see what another teacher has done with this topic in the article "Cycling Through Plants" (Cavallo 2005). Both are available online at *www.nsta.org*.

For very young children, you may want to focus on just one or two of the possible variables. For example, planting seeds in cups with too much water, no water, and just enough water can give important information to your students. They can also learn to control variables in this simpler type of investigation.

USING THE STORIES WITH GRADES 5–8

For older children, I also suggest giving the probe "Needs of Seeds" (Keeley, Eberle, and Tugel 2007) as mentioned in the previous section. Formative assessment is

so important that starting to teach without knowing what your students think about a concept is like heading out to sea in a rowboat without oars or a map and compass. It can also generate a discussion, especially if there is a lack of agreement on the part of the students. From this discussion, almost certainly, suggestions for testing ideas will result.

Many of the ideas listed above will be pertinent to your approach to using the story with older children. They will be able to identify the variables involved in setting up the test to see if various things are necessary for the seeds to germinate. It is always advisable not to assume too much about how facile students are in designing investigations. They may well need your help in identifying variables that can be modified or kept the same. They may also need a refresher on designing investigations, which includes controlling variables. If you ask small groups of students to design a way to answer the questions, they can present their designs to the class for critique.

I believe that it is important to help students to learn to give constructive criticism without being insulting. I usually ask them to begin by telling the group what they like about the design and then saying something like, "However, I have a problem with the fact that there is nothing in the design to keep track of the amounts of water/soil/fertilizer you will be using." This technique may take the sting out of criticism and make the recipients less defensive and more receptive. As I have stated in the early chapters, we need to help our students to be communicators and be helpful to each other. We do not spend enough energy in teaching them how to "talk science" and to listen to others.

NSTA members have free access to the journal article, "Bean Plants: a Growth Experience" (West 2004) at *www.nsta.org*. It is always helpful to find out what others have done with similar units. In this article, West describes how she carried out an investigation about the growth of bean plants and addressed her students' misconceptions. She also includes a very useful rubric she designed for assessing the growth of her students' knowledge about carrying out the investigation.

Students who are older may also want to make the study a bit more complex by adding more questions regarding quantities of materials and graphing results. For example:

- If a little water is good, is more better?
- Do seeds need deep soil to germinate?
- How much warmth do seeds need?
- Is any soil used up in the process of germination?

As in any activity designed to modify preconceptions that do not agree with the currently accepted ideas in science, your results should be fairly convincing that seeds need only three things to germinate and that fertilizer has no part of determining if a seed germinates. If, however, some students still hold fast to their previous ideas, be satisfied that you have helped to add another beam in the scaffolding that will someday result in their understanding the needs of seeds.

related NSTA Press BOOks and NSTA JOURNAL articles

Keeley, P., 2005. *Science curriculum topic study: Bridging the gap between standards and practice.* Thousand Oaks, CA: Corwin Press.

Keeley, P., F. Eberle, and J. Tugel.2007. *Uncovering student ideas in science, volume 2: 25 more formative assessment probes.* Arlington, VA: NSTA Press.

Keeley, P., F. Eberle, and C. Dorsey. 2008. *Uncovering student ideas in science, volume 3: Another 25 formative assessment probes.* Arlington, VA: NSTA Press.

Keeley, P., and J. Tugel. 2009. *Uncovering student ideas in science, volume 4: 25 new formative assessment probes.* Arlington, VA: NSTA Press.

Konicek-Moran, R. 2008. *Everyday Science Mysteries: Stories for inquiry-based science teaching.* Arlington, VA: NSTA Press.

Konicek-Moran, R. 2009. *More everyday science mysteries: Stories for inquiry-based science teaching.* Arlington, VA: NSTA Press.

Konicek-Moran, R. 2010. *Even More Everyday Science Mysteries: Stories for inquiry-based science teaching.* Arlington, VA: NSTA Press.

references

American Association for the Advancement of Science (AAAS).1993. *Benchmarks for science literacy.* New York: Oxford University Press.

Ashbrook, P. 2005. Teaching through trade books: What happens to seeds? *Science & Children* 42 (8): 18–20.

Cavallo, A. 2005. Cycling through plants. *Science & Children* 42 (8): 22–27.

Botanical Online SL. 2010. The germination of the bean. *www.botanical-online.com/animation4.htm*

Driver, R., A. Squires, P. Rushworth, and V. Wood-Robinson. 1994. *Making sense of secondary science: Research into children's ideas.* London and New York: Routledge Falmer.

Keeley, P., F. Eberle, and J. Tugel.2007. *Uncovering student ideas in science, volume 2: 25 more formative assessment probes.* Arlington, VA: NSTA Press.

Konicek-Moran, R. 2008. *Everyday Science Mysteries: Stories for inquiry-based science teaching.* Arlington, VA: NSTA Press.

Konicek-Moran, R. 2009. *More everyday science mysteries: Stories for inquiry-based science teaching.* Arlington, VA: NSTA Press.

National Research Council (NRC). 1996. *National science education standards.* Washington, DC: National Academy Press.

West, D. 2004. Bean Plants: A growth experience. *Science Scope* 27 (7): 44–47.

CHAPTER 12
REACTION TIME

Fouad and his brother Abdul were playing a game of "hand slap," which they had learned in their native land. In this game, sitting opposite one another on the floor, one player places his hands face up on his lap and the other lays his hands face down on top of the first player's hands, palm to palm. The one whose hands are on the bottom tries to take his hands away fast and slap the hands of the other before they are pulled back. This requires a great deal of quickness to feel the other's hands move and pull back so that your hands cannot be slapped. And it takes speed to remove your hands quickly without being noticed and to slap the hands of the other person. People often call this *reaction time*.

Fouad seemed to win at this game more than his brother and liked to brag a little about it.

"I have such a good jump on you, Abdul! I know when you are going to move your hands before you do yourself. I have excellent reaction time."

"Well, maybe you do win more than I do but I think I have really good reaction time too."

"Well, you sure don't show it in the game," said Fouad. "If you can't win at hand slap, why do you think you have good reaction time?"

"Because I am quick at other games. You only beat me at this one!"

"I saw something in a magazine that will help us to decide who is faster!" said Fouad. "Do you want to give it a try?"

"Sure, as long as it's fair," said Abdul. "What is the test?"

"You have to get a meterstick first. One person holds it at the top and the other person opens his first finger and thumb apart at the bottom of the stick. The person who is holding the stick lets it go and the other person pinches his fingers together as soon as he notices the stick falling to try to stop it. How far down the stick travels measures how fast the other person reacted. The less the stick falls, the quicker you grab it and the better your reaction time."

"That sounds cool! I think I can think of all kinds of reaction times we can measure with that test. Let's get a meterstick and do it! But remember we have to make sure it's fair."

"Okay, we'll make sure it's fair but what do you mean by different kinds of reaction time?"

"You'll see," said Abdul. "Maybe you won't be the best at all of the different kinds."

PURPOSE

This story provides the opportunity for students to learn about how our nervous system responds to stimuli and to find out how our nervous system works.

RELATED CONCEPTS

- Nervous system
- Simple reaction time
- Stimuli
- Responses
- Averages

DON'T BE SURPRISED

Students may have the impression that all kinds of reaction time responses can be improved through practice. They may believe also that every kind of stimulus produces the same kind of reaction time. They are not usually aware of the importance of the central nervous system's role in determining reaction time.

CONTENT BACKGROUND

Reaction time is the interval of time between the presentation of a stimulus and the resulting behavioral reaction in any living organism. In sports, reaction time is crucial. Take for example, baseball: A major-league league pitcher stands about 60 feet and 6 inches from the batter who is armed with a bat. Many pitchers can throw a baseball at 95 miles per hour or faster. That means the ball reaches the batter in approximately 0.4 seconds. The batter has about .04 seconds to decide where or even whether to hit it at all. That requires very fast reaction time. Most batters have trouble doing so, and hit the ball less than 3 times in 10. Any more than that and the batter is considered to be an excellent hitter. In 1941, Ted Williams was one of only seven hitters since 1903 to hit over .400 (four hits out of every 10 times at bat). Wildlife experts tell us that predators like the cheetah and the lion are successful only three times in ten. Life is tough in the wild *and* on the baseball diamond!

This story has to do with the reactions of the nervous system. In many animals there are two parts to the nervous system, the central and the peripheral. The *central nervous system* is made up of the brain, the spinal cord, and the retina. The *peripheral nervous system* is made up mainly of sensory neurons that are connected to the central nervous system. Comprised of cells called *neurons*, the nervous system sends signals from neuron to neuron via electrochemical waves that travel along very thin fibers called *axons*. Chemicals called neurotransmitters are released at the junctions of these cells, called *synapses*, which allow the signal to continue on to their destinations via the next set of axons.

For example, when a person reacts to a visual stimulus, many things have to happen before that reaction takes place. The stimulus (in the form of light particles or waves) enters the eye and then goes, via the neurons or *neural pathways* to the brain where it is processed and a response chosen. The response or directions for physical reaction is then sent over the myriad of neurons to the spinal cord. The spinal cord sends the impulse on to the proper muscles, telling them how to react. Clearly, then, there are a lot of transfers of this electrical-chemical information, but it is all happening at super speed.

But hey, sometimes even *that* is too slow for your own good! Suppose someone throws something and it comes your way—maybe even at your face. A *reflex* action takes place in this case and the message goes directly to the spinal cord, bypassing the brain. You duck or blink your eyes to avoid contact. Similarly, you may touch something hot. The receptor in your skin sends the message "HOT!!" to your spinal cord that immediately causes you to pull back. Your brain doesn't get a chance to "think" about it. Good thing too because it can often save you from serious injury.

But now back to the story and reaction time. Reaction time does involve the brain. In certain situations, known as *simple reactions*, where a person reacts to a single stimulus and responds with a conscious action, a sensory organ or nerve (using sight, hearing, or touch) is involved and the neurons send the message to the brain, then to the spinal cord and on to the muscles. It is "hard wired" and does not usually improve beyond a certain point even with practice. It also hits its peak in the teens and then usually goes downhill progressively with age. I suspect that is why some professional athletes are "over the hill" by age 35. That 95-mile-per-hour fastball begins to look like an aspirin tablet and is, more often than not, a strike, either looking or swinging.

The internet is loaded with different reaction time tests that you can try. The reaction time tester referred to in the story is a simple one and uses only a meter or yardstick found in most households. You and some of your more interested students might like to read a literature review by Robert Kosinski of Clemson University at *http://biology.clemson.edu/bpc/bp/Lab/110/reaction.htm*. This way your students can see that scientists at major universities are doing research on the same topic as they are. This should end doubts that they too, can be real scientists.

There is also a different type of test that is called a "go/no-go" test in which the subject must decide between two stimuli and either react or not react depending upon the directions. For example, the subject may have to refrain from responding to a certain stimulus and respond to another. Therefore, the brain must process more than one symbol and then tell the subject how to respond (or not). This of course takes much more reaction time and is not mentioned in the story because it is much too complicated to measure in a classroom with everyday materials. However, the test is available online at *http://cognitivefun.net/test/17*.

related Ideas From the National Science education standards (NrC 1996)

K–4: Characteristics of Organisms

- The behavior of individual organisms is influenced by internal cues (such as hunger) and by external cues (such as a change in the environment). Humans and other organisms have senses that help them detect internal and external cues.

5–8: Regulation and Behavior

- Regulation of an organism's internal environment involves sensing the internal environment and changing physiological activities to keep conditions within the range required to survive.
- Behavior is one kind of response an organism can make to an internal or environmental stimulus. A behavioral response requires coordination and communication at many levels, including cells, organ systems, and whole organisms. A behavioral response is a set of actions determined in part by heredity and in part from experience.

5–8: Structure and Function in Living Systems

- Specialized cells perform specialized functions in multicellular organisms. Groups of specialized cells cooperate to form a tissue, such as a muscle. Different tissues are in turn grouped together to form larger functional units, called organs. Each type of cell, tissue, and organ has a distinct structure and set of functions that serve the organism as a whole.
- The human organism has systems for digestion, respiration, reproduction, circulation, excretion, movement, control, and coordination, and for protection from disease. These systems interact with one another.

RELATED IDEAS FROM BENCHMARKS FOR SCIENCE LITERACY (AAAS 1993)

K–2: The Human Organism

- Senses can warn individuals about danger; muscles help them to fight, hide, or get out of danger.
- The brain enables human beings to think and sends messages to other body parts to help them work properly.

3–5: The Human Organism

- The brain gets signals from all parts of the body telling what is going on there. The brain also sends signals to parts of the body to influence what they do.

6–8: The Human Organism

- Organs and organ systems are composed of cells and help to provide all cells with basic needs.
- Interactions among the senses, nerves, and brain make possible the learning that enables human beings to cope with changes in their environment.

USING THE STORIES WITH GRADES K–4

What do you think your kids are going to want to do first? The hand slap game, right? It is described pretty thoroughly in the story. The children can see that it is stealth and quickness that win in this game. Ask them to analyze the game and try to decide what they are each doing to fool the other person. They should be able to see that it is how quickly the person on top can sense the movement of the other person's hands that make it possible to avoid the slap. Players usually set a number such as 10 tries to see who wins the game and then the partners switch roles and try again. Scoring can be at the discretion of the players. Questions usually arise, such as:

- Do you get better with practice?
- Do you develop a strategy to fool the other person?
- Do you think your reaction time gets any faster or do you just figure out your opponent's moves? (Note: Actual response time is hard wired but strategies do develop with practice.)
- Are you able to do better with an older person? A younger person?

Being the teacher, you'll have to accept challenges about the last question and prepare to lose since reaction time does get slower as you age. But, you can accept defeat in the interest of education, right?

With children in second grade and above you should feel comfortable in letting them try to measure reaction that will result in numerical data. The only piece of equipment you will need is a ruler, yardstick, or meterstick. Again, this is described pretty thoroughly in the story. The stick should be held so that the bottom is in line with the thumb and finger of the "catcher." The "dropper" drops the stick without warning and the catchers pinch thumb and finger together as quickly as possible. The distance the stick fell can be measured by seeing where the middle of the thumb or finger is on the stick. This distance can be recorded in the science notebooks and analyzed to see if there are changes as trials progress.

Again, the reaction time will not change but catching strategies may make a difference as practice continues. Fatigue, loss of interest, or boredom can have an affect on the reaction time as well as drugs, even coffee. I have already mentioned age as a factor. Reaction time shortens from infancy until the 20s and then begins to lengthen slowly until about the 50s and 60s. After that, the response lengthens more quickly with advancing age. At my age (79), I guess that leaves me pretty much close to coming in last place! And, sorry women, but females seem to be a bit slower than males on average (Der and Deary 2006). We have evidence that auditory stimuli result in shorter reaction times than visual stimuli (Brebner and Welford 1980). This seems to be because it takes less time for an auditory stimulus to reach the brain for processing than a visual one. It also seems that in computer mouse tests, left-handed people do better than right-handed subjects (Peters and Ivanoff 1999). This makes sense, as left-handed people often have dominant right brain hemispheres, which is where spatial tasks are performed.

There are still many things that are unknown about the neural networks that form reaction time. Students have been known to raise some new and interesting questions. These may be too sophisticated for very young children but if you have a class that seems able to handle more difficult tasks, read the next section and see if your children could do some of them.

USING THE STORIES WITH GRADES 5–8

It should not surprise anyone if middle school students wanted to begin working with the hand slap game just as much as the elementary age children. They may wonder what Abdul is talking about when he says that Fouad may not be best at all kinds of reaction time tasks. Once they have tried the meterstick drop, my experience is that the students will come up with many more questions than you will. Each is usually investigable and deserves to be tested. Some will be:

- How will an auditory stimulus affect reaction time in a blindfolded subject?
- How will a touch stimulus affect reaction time in a blindfolded subject?
- How does age affect reaction time?
- Does age affect reaction time in all tasks?
- Does gender affect reaction time?
- Does handedness affect reaction time?

Fouad and Abdul stressed that the tests be fair and so they should. You can help students make sure that all variables are maintained in a proper way regardless of the question being investigated. Data collected in all tests should be recorded, both in individual notebooks and as a class. It is wise to do this anonymously so that some less quick students do not feel embarrassed. You can also think about doing a larger test including more than one class or testing across ages. Graphing can be introduced as a way to use a pictorial summary of data. You will probably end up with a bell shaped curve, with atypical results at the extremes and the more typical results in the middle. This also is a good introduction to statistics. Students can use statistics to predict the most likely results for individuals who are tested after the graph is made.

You will undoubtedly have to put an end to the investigations because the students will want to continue to test new ideas and new groups of individuals. But, all in all, I think I can promise you that your class will involve themselves in real inquiry and data collection and that it will be worthwhile for many pedagogical reasons.

For individuals who want to be challenged, you can use your Internet search engine to find more reaction time tests online that will be both fun and informative. Please take time to do this since you will be able to use various stimuli and tests like "go/no go" time reaction in an entertaining way.

related NSTA Press Books and NSTA Journal articles

Keeley, P, 2005. *Science curriculum topic study: Bridging the gap between standards and practice.* Thousand Oaks, CA: Corwin Press.

Keeley, P., F. Eberle, and J. Tugel. 2007. *Uncovering student ideas in science, volume 2: 25 more formative assessment probes.* Arlington, VA: NSTA Press.

Keeley, P., F. Eberle, and C. Dorsey. 2008. *Uncovering student ideas in science, volume 3: Another 25 formative assessment probes.* Arlington, VA: NSTA Press.

Keeley, P., and J. Tugel. 2009. *Uncovering student ideas in science, volume 4: 25 new formative assessment probes.* Arlington, VA: NSTA Press.

Konicek-Moran, R. 2008. *Everyday Science Mysteries: Stories for inquiry-based science teaching.* Arlington, VA: NSTA Press.

Konicek-Moran, R. 2009. *More everyday science mysteries: Stories for inquiry-based science teaching.* Arlington, VA: NSTA Press.

Konicek-Moran, R. 2010. *Even More Everyday Science Mysteries: Stories for inquiry-based science teaching.* Arlington, VA: NSTA Press.

references

American Association for the Advancement of Science (AAAS).1993. *Benchmarks for science literacy.* New York: Oxford University Press.

Brebner, J. T., and A. T. Welford. 1980. Introduction: An historical background sketch. *In Reaction Times,* ed. A. T. Welford, 1–23. New York: Academic Press.

Cognitive Fun. 2008. *http://cognitivefun.net/test/17*

Der, G., and I. J. Deary. 2006. Age and sex differences in reaction time in adulthood: Results from the

United Kingdom health and lifestyle survey. *Psychology and Aging* 21 (1): 62–73.

Peters, M., and J. Ivanoff. 1999. Performance asymmetries in computer mouse control of right-handers, and left-handers with left- and right-handed mouse experience. *Journal of Motor Behavior* 31 (1): 86–89.

National Research Council (NRC). 1996. *National science education standards.* Washington, DC: National Academy Press.

CHAPTER 13
SEEDLINGS IN A JAR

Sara and Ina were having an argument. Well, maybe not an argument; more like a disagreement about a science topic. Ina and Sara wanted to be scientists when they grew up. They had watched a science video about how a plant gets most of the stuff that makes up its stems and leaves—what the scientists in the video called *mass*—from the air, as it grows. Sara wanted to know how this happened. She couldn't believe that air could hold anything that could build a plant.

She thought and thought about it, and the next day, she suggested to Ina that they try an experiment and grow a plant in a closed jar so that she could prove that plants didn't get their mass from the air.

"There isn't enough air in this big pickle jar to make leaves and stems for a whole plant so if they grow, we know that the stuff that makes the plant comes from the soil or the water."

"Well, I'm not even sure we can grow plants in a jar, and not even sure we can sprout seeds in a jar with so little air," said Ina.

"That's just the point," Sara said. "If they need air to make them get bigger, they won't grow in a jar. So if they *do* grow—which they won't, I'm sure!—we can prove I am right."

"Okay, but I want to have some control over this experiment because I don't think it will prove anything by just planting seeds in a closed jar," said Ina. "I want to weigh everything we use so we can tell if anything is missing when it is over." Ina had done scientific experiments before, and she knew about weighing ingredients and keeping track of them in her science notebook.

"Fair enough," said Sara. "It won't make any difference anyway so if you want to go to all of the trouble of weighing everything, be my guest. Anyway I'll bet the soil will weigh less, because *that* is where the plant gets its mass."

So they did weigh the jar and cover, the soil they put in, the water they added and, of course, the seeds they planted in the soil. Ina kept accurate records in her science notebook about everything they did. She even recorded the weight of the whole thing after it was put together and sealed up.

Several weeks went by. The covered jar got very misty inside and, to their surprise, the seeds germinated and the seedlings grew up almost to the top of the jar.

"There, you see!" said Sara. "The jar was closed as tight as can be, so no air leaked in or out and the plants grew just fine. The mass must have come from the soil or the water."

"Not so fast," said Ina. "We have some weighing to do."

"What in the world for?" asked Sara. "We just proved that air had nothing to do with the seeds sprouting or the plants growing, didn't we?"

"Not to me we didn't," said Ina. "I want to weigh everything again."

"Suit yourself, Ina," said Sara. "It's a waste of time. It will weigh less because the soil and water got used up."

Sara and Ina weighed the whole set up and both said, "OMG!"

Neither girl could believe their eyes. "Weigh it again," said Sara. "There has to be something wrong!"

Ina weighed the whole thing and got the same answer. Now neither girl knew what had happened.

"Let's weigh everything separately and see what it adds up to," said Sara.

They did and it all added up to the same number. They weighed the soil, which was wet with moisture, the jar, the lid, and the plants, which by now were quite large.

"Okay, now what?" said Sara.

"I think we really need to do this again," said Ina.

Purpose

This story really leaves us hanging! It's a great one for real investigation! There are two purposes of the story: the first is to investigate closed systems. Anything that happens inside uses up only the materials in the jar, because there is no access to the outside world. The second purpose is for students to understand that air has mass and contains the materials that plants need to make their food and build their structure. I think a further purpose of the story is to allow the students to be puzzled by the outcome, and learn from this puzzlement how to try to find out how things happen.

related concepts

- Systems
- Closed systems
- Photosynthesis
- Germination
- Contents of the atmosphere
- Experimental design

DON'T BE surprised

Your students will probably be very surprised at the outcome of their investigation. It would seem logical that the system would gain weight since the plants are growing inside the jar and they must weigh something! They are probably not aware of at least two things that are important, the implications of a closed system and the concept of air as having mass.

CONTENT BacKGROUND

This story is about systems more than it is about seeds and plant growth, but you will want some information about the latter, so I will provide that in this section. But first, let us look at the concept of systems.

A *system* is a set of individual entities that interact and influence each other in the performance of a given task. The entities can be objects, people, internal organs, buses, plants, or dozens of other things that somehow interact in a meaningful way. The American Association for the Advancement of Science (1993) believes the idea of systems to be one of the most important overarching concepts in learning and strongly urges educators to make the concept of systems a central part of all subjects taught in schools. After all, we are surrounded by systems, we live within systems and anything we teach or study is encased somehow in some system, somewhere.

Students can apply the idea of systems to any topic and, most importantly, could see the relationships among the various topics they study. We so often complain that students do not see the big picture, but learn things in isolated units and then fail to apply them across the curriculum. If systems were used as a unifying theme across all subjects, I believe that there would be more complete understanding of how everything in our everyday lives is related. In other words, they would see how the ideas involved in a transportation or political system related to how things work in an ecosystem or even a digestive system. Any change in any system affects everything else in that system.

Systems are said to be **closed** or **open**. Actually there is a third type, an **isolated** system, but it has little importance here. A *closed system* such as the one in the story can *exchange energy but not matter*. The matter that is in the jar remains constant and cannot interact with the outside environment. However, sunlight can get through the glass and provide energy for photosynthesis. A hot liquid in a container, like coffee in a thermos, is another example of a closed system. Thermal energy can pass through the container walls so that eventually the hot liquid reaches equilibrium with its environment, but the coffee sealed inside can neither take in nor give out any of its substance.

An *open system* allows both matter and energy to be exchanged with the environment. Our Earth is an example of an open system, where everything can react with everything else. Matter and energy are being exchanged all of the time. If the jar in the story had not been sealed off, it would have been classified as an open system since it could exchange matter and energy with anything outside the jar.

With the idea of the seeds germinating and growing in what Sara considered an impossible growing environment, we add a different content twist to the concept of a closed system. Sara was under the influence of the popular misconception that air has no mass and therefore cannot possibly be part of the materials involved in the production of the plant tissues, or the *mass*, of growing plants. She believed that the plant mass comes from the soil and water, through the roots. She also believed that there was not enough air in the jar to grow anything.

In the 1600s Jan Baptista Van Helmont planted a willow sapling in 200 pounds of soil, which after five years gained 164 pounds. (Obviously, he kept track of weights in his science notebook!). The soil lost only 57 grams of soil (2 ounces) during that time. Realizing that the loss in soil mass could not be responsible for the gigantic change in mass, he assumed it was the rainwater he had added that made up the difference in mass. Although he was aware of the carbon dioxide in the atmosphere, he could not believe that it could be responsible for the 164 pounds of additional weight. Here is an example of a misconception held by a scientist who did not have enough information to reach a conclusion which we now know is acceptable. This conclusion is that the carbon from the carbon dioxide in the air and water obtained through the roots can provide enough mass through photosynthesis to account for the gain in mass of the plants that grow on this Earth. The carbon and hydrogen in the presence of solar energy and chlorophyll are used to make sugars, which bond together to make starch and finally cellulose to form the cell walls of the cells from which plant tissues are made. Byproducts are newly formed water and oxygen, which are released into the atmosphere. Thus, the girls noticed the droplets of water on the jar's inside surface, which showed

that the water had condensed on the glass. There was enough carbon dioxide in the jar to provide the carbon necessary to allow plant growth. Since this was a closed system, mass was conserved since the only material available to the seed and subsequent plants was contained in the jar and the system should not gain any weight even though the appearance inside the jar changed considerably. Since the jar was transparent, solar energy could enter the system from the outside. Excess heat, if there was any could also be released through the glass and so energy could be exchanged. This permits us to use our definition of a closed system very well. The chemical formula that includes the formation of the new water molecules is $6\ CO_2 + 12\ H_2O$ + energy and chlorophyll $-> C_6H_{12}O_6 + 6\ O_2 + 6\ H_2O$

Seeds need warmth and moisture in order to germinate. A few seeds, such as lettuce seeds also need light to activate the processes involved in germination. Some maple seeds can only last for a few weeks before they die if they do not germinate. Other seeds, such as the Lotus can remain viable for thousands of years. But it is the water and warmth that allow the seed to germinate. Water enters the seed, softens the seed coat and allows the chemical processes necessary for germination to proceed. Gibberellic acid is released and begins a series of rather complicated chemical and genetic steps that result in a dormant seed becoming a full-fledged plant. This plant is capable of producing more seeds so that the life cycle is completed. In the jar in the story, the girls provided the water and the growth medium, the soil. The room temperature and the Sun provided the warmth and that is all the seeds needed to germinate. The sunlight penetrating the glass jar walls provided the energy to cause the germinated seeds to grow into seedlings and then plants. As you may notice from the formula for photosynthesis, oxygen and water are also produced in the process and continue to provide necessary elements for plant growth. Plants need oxygen in order to respire (break down nutrients to produce useful energy) and water. The jar is a little microcosm of the outside world and might continue to exist for some time barring some changes inside that might inhibit the processes and stop the cycle. In the open system of the outside world, pests, disease, climate change, and a host of other problems may affect plants in the ecosystem. Yet, the open system of our Earth has survived for millions of years, changing in ways that have kept succession alive. We are now facing the prospects of what we humans have done to the ecosystem at large that may have dire consequences for living things on the planet.

related ideas from the national science education standards (Nrc 1996)

K–12: Unifying Concepts and Processes—Systems

- The natural and designed world is now too large and complex to comprehend all at once. Scientists define small portions for the convenience of investigation. These portions are referred to as *systems*.

- A system is an organized group of related objects or components that form a whole. Systems have boundaries, components, resources, flow (input and output), and feedback.
- Within systems, interactions between components occur.
- Systems at different levels of organization can manifest different properties and functions.
- Thinking in terms of simple systems encompasses subsystems as well as identifying the structure and function of systems, feedback and equilibrium and identifying the distinction between open and closed systems.
- Understanding the regularities in systems can develop understanding of basic laws, theories, and models that explain the world.

K–4: The Characteristics of Organisms
- Organisms have basic needs. For example animals need air, water and food; plants require air, water, nutrients and light. Organisms can survive only in environments in which their needs can be met.
- The world has many different environments and distinct environments support the life of different types of organisms.
- Each plant or animal has different structures that serve different functions in growth, survival, and reproduction.

K–4: Life Cycles of Organisms
- Plants and animals have life cycles that include being born, developing into adults, reproducing, and eventually dying. The details of this life cycle are different for different organisms.
- Plants and animals closely resemble their parents.

5–8: Structure and Function in Living Systems
- Living systems at all levels of organization demonstrate the complementary nature of structure and function. Important levels of organization for structure and function include cells, organs, tissues, organ systems, whole organisms, and ecosystems.

5–8: Reproduction and Heredity
- Reproduction is a characteristic of all living systems because no individual organism lives forever. Reproduction is essential to the continuation of every species. Some organisms reproduce asexually. Other organisms reproduce sexually.

5–8: Diversity and Adaptations of Organisms
- Millions of species of animals, plants, and microorganisms are alive today. Although different species might look dissimilar, the unit among organisms becomes apparent from an analysis of internal structures, the similarity of their chemical processes and the evidence of common ancestry.

related ideas from Benchmarks for science Literacy (aaas 1993)

K–2: Systems

- Most things are made of parts.
- Something may not work if some of the parts are missing.
- When parts are put together, they can do things that they couldn't do by themselves.

3–5: Systems

- In something that consists of many parts, the parts usually influence one another.

6–8: Systems

- A system can include processes as well as things.
- Thinking about things as systems means looking for how every part relates to others. The output from one part of a system (which can include material energy or information) can become the input to other parts. Such feedback can serve to control what goes on in the system as a whole.
- Any system is usually connected to other systems, both internally and externally. Thus a system may be thought of as containing subsystems and as being a subsystem of a larger system.
- Some portion of the output of a system may be fed back to the system's input.
- Systems are defined by placing boundaries around collections of interrelated things to make them easier to study. Regardless of where the boundaries are placed, a system still interacts with its surrounding environment. Therefore, when studying a system it is important to keep track of what enters or leaves the system.

K–2: Diversity of Life

- Some animals and plants are alike in the way they look and in the things they do, and others are very different from one another.
- Plants and animals have features that help them live in different environments.

K–2: The Physical Setting

- Things can be done to materials to change some of their properties, but not all materials respond the same way to what is done to them.

3–5: Diversity of Life

- A great variety of kinds of living things can be sorted into groups in many ways using various features to decide which things belong in which group.
- Features used for grouping depend on the purpose of the grouping.

6–8: Diversity of Life

- Animals and plants have a great variety of body plans and internal structures that contribute to their being able to make or find food and reproduce.
- For sexually reproducing organisms, a species comprises all organisms that can mate with one another to produce fertile offspring.

USING THE STORIES WITH GRADES K–4

Young children can examine their toys and things that they build or use to identify the parts that make up what we call systems. According to the standards, they should realize that toys are made of parts and that if some parts are missing, the toy does not work correctly. This can be an important introduction to the concept of systems. A bicycle cannot operate without a wheel and an electronic toy cannot operate without a source of power such as batteries or a wind-up spring. Looking at toys and identifying their parts can also be instructive for children as they develop the idea of systems. It is important that young children begin their understanding of systems by realizing that the parts of systems interact with each other in ways that make the system work properly.

When I was a young boy I became enamored with locks and to my parents' chagrin I managed to take apart most of the locks on the doors of our house. It seemed like I always had parts left over when I reassembled them. Of course they didn't work and we had to start all over again. But in doing so, I learned more about how locks work than I would have in any other way. I wanted to believe that it made no difference if one little part was left out of the lock system. But, of course, it did. Helping children to see that successful systems have to consist of all of their parts is one of the most important things they can learn. The application of this idea to the larger examples of systems in the world will prove invaluable.

You might try giving the probe "Is it a system?" from *Uncovering Student Ideas in Science, Volume 4* (Keeley and Tugel 2009). This probe will provide you with more information about your students' knowledge about systems. Students are asked to choose from a list, those things that they think are systems, and then explain their reasoning. In all of these activities, the important thing is the discussion that follows. A teacher learns more about how and what their students learn by listening than by talking. You may hear students holding opposite points of view and not even realizing it. With careful probing you can help them to see what they are saying is in conflict. I once had a student who was in the middle of an explanation and stopped and said, "now that *I heard what I just said,* I can see that I am heading in the wrong direction." Hearings oneself speak out loud is often a better key to understanding what we are thinking than just pondering it silently. Additionally, writing it or drawing it can also add information that aids in understanding, especially, if one is writing for oneself as the audience. The writer can be more honest that way.

Your students may want to try the activity in the story. Young children may not be ready to use the skills necessary to carry out the activity. Older elementary students who can use a balance or scale and follow the story line carefully can repeat the activity and after predicting the outcome, can find the seemingly amazing outcome for themselves. They may want to see how long the microcosm in the jar will perpetuate itself. If you help them to use plants that will not grow too tall too quickly, the chances are that it might go on for some time.

USING THE STORIES WITH GRADES 5-8

Middle school children will be just as surprised as the younger students with the results and you might be too. Don't be surprised if you expected the jar of plants to weigh more. It almost seems counterintuitive that the plants could grow from the seeds and still not add any weight to the system. This particular system however, is the secret to the result. Try to focus on the system being closed and appeal to their knowledge that nothing in and nothing out yields no changes.

The inspiration for the story came from the probe, "Seedlings in a Jar," in Keeley and Tugel's book. (Keeley and Tugel 2009). Although giving this probe may seem redundant since it follows the story so closely, it could be used as a summative assessment. You would need to pay closer attention to the part of the probe where the students are required to elucidate in their own words why the system did not gain weight.

When the students decide to replicate the story, you may want to talk with them about systems and have them find as many different kinds of systems that they can think of in their collective experiences. This particular moment in the process is the best time to introduce the vocabulary since the words will be in context. Older students, being more facile with balances will want to weigh the substances before and after and check the results several times. I have found that their disbelief based on prior conceptions, is strong and sometimes hard to shake. Prior conceptions held for a long time are difficult to change with one activity no matter how strong the evidence may seem to us. If you do not get all of your students to understand the concepts involved here, do not be discouraged. Instead, congratulate yourself on having built one more plank in the scaffolding that leads to final understanding. Believing that the air in the jar was sufficient to contain the building blocks of the plant tissue is completely ridiculous to students who believe that air has no mass. If you are planning to follow up this story with photosynthesis, I believe that you will find the students more receptive to the ideas involved because they have actually seen results that point to the carbon dioxide in the air as the source of carbon that will be the backbone of the starch and sugar molecules and eventually the cellulose.

If you would like to see an interview of a young student by an interviewer who uncovers his lack of understanding about photosynthesis, log onto the following: *www.learner.org/resources/series26.html* and watch the video "Lessons From Thin Air." It will give you an idea how difficult it is to change student conceptions about air and mass.

RELATED NSTA PRESS BOOKS AND NSTA JOURNAL ARTICLES

Keeley, P, 2005. *Science curriculum topic study: Bridging the gap between standards and practice.* Thousand Oaks, CA: Corwin Press.

Keeley, P., F. Eberle, and L. Farrin. 2005. *Uncovering student ideas in science, volume 1: 25 formative assessment probes.* Arlington, VA: NSTA Press.

Keeley, P., F. Eberle, and J. Tugel.2007. *Uncovering student ideas in science, volume 2: 25 more formative assessment probes.* Arlington, VA: NSTA Press.

Keeley, P., F. Eberle, and C. Dorsey. 2008. *Uncovering student ideas in science, volume 3: Another 25 formative assessment probes.* Arlington, VA: NSTA Press.

Keeley, P., and J. Tugel. 2009. *Uncovering student ideas in science, volume 4: 25 new formative assessment probes.* Arlington, VA: NSTA Press.

Konicek-Moran, R. 2008. *Everyday Science Mysteries: Stories for inquiry-based science teaching.* Arlington, VA: NSTA Press.

Konicek-Moran, R. 2009. *More everyday science mysteries: Stories for inquiry-based science teaching.* Arlington, VA: NSTA Press.

Konicek-Moran, R. 2010. *Even More Everyday Science Mysteries: Stories for inquiry-based science teaching.* Arlington, VA: NSTA Press.

REFERENCES

Annenberg Media. 2009. Lessons from thin air. Minds of Our Own. *www.learner. org/resources/series26.html*

Keeley, P., and J. Tugel. 2009. *Uncovering student ideas in science: 25 formative assessment probes, Volume 4.* Arlington, VA: NSTA Press.

PHYSICAL SCIENCES

Core Concepts	Sweet Talk	Cooling Off	Party Meltdown	The Crooked Swing	The Cookie Dilemma	Stuck!
Melting	X		X			
Dissolving	X					
Temperature	X	X	X			X
State Change	X	X	X			
Mixtures	X	X			X	
Solvent	X					
Solute	X					
Heat	X	X	X		X	X
Thermal Energy	X	X	X		X	X
Thermometers		X				
Calorimeters		X				
Conduction		X	X			X
Equilibrium		X	X			
Thermal Conductivity			X			X
Periodic Motion				X		
Problem Solving	X	X	X	X	X	X
Model Building		X		X		
Problem Analysis	X	X	X	X	X	X
Creativity				X	X	
Systems	X	X	X	X	X	X
Chemical Reactions					X	
Friction						X
Static Friction						X
Kinetic Friction						X
Net Force						X
Force						X
Interaction	X	X	X	X	X	X

CHAPTER 14
SWEET TALK

You may remember Caroline and Lisa, the sisters in the story "Iced Tea" in *More Everyday Mysteries*. Let's go back to their porch and listen to another conversation they had about adding sugar to iced tea. It provides a look at another everyday science mystery and at two words that are very confusing to a lot of people.

Caroline was stirring her iced tea madly.

"What are you doing?" Lisa asked. She was a lot older than Caroline, but she always stopped to check in with her little sister. She sat on the porch swing with some nail polish and began to paint her nails.

"I'm trying to make this sugar melt!" Caroline answered. "It's a lot harder with iced tea than when you put sugar in hot tea!"

"Actually you aren't *melting* anything," said Lisa, "You are *dissolving* the sugar—or at least you're trying to." Lisa loved her science classes in the high school.

"Melting, dissolving … what's the difference?"

"If you knew the difference you probably wouldn't be doing what you are doing!" said Lisa. "What kinds of things do you know that melt?"

"Ice cream, ice cubes, candy in my mouth, sugar in hot tea, chocolate, and… well, lots of things!"

"Well, that list is *almost* right." Lisa answered. She worked on her nails for a few moments. "Well, then, what kinds of things do you know that dissolve?" she asked, after a while.

"Aren't they the same?" asked Caroline, looking glumly at the bits of sugar swirling at the bottom of her glass.

"Nope. Okay, I'll give you a hint. I think you are clever enough to know which things in your life dissolve and which things melt."

"Better try me and see," said Caroline. "Maybe it takes more than being clever."

"Think of the difference between putting sugar into iced tea and watching your ice cream cone dribble all over the front of you. The hint is what is *missing* in one case but *present* in the other?"

"If I can figure that riddle out, can get my iced tea sweet?" asked Caroline.

"I think so but, then again, maybe not. Want to bet on it?" asked Lisa. And she gave her sister a sly smile.

PURPOSE

This story's purpose is to acquaint students with the very different physical processes of melting and dissolving, which will help further their understanding of various changes of states.

RELATED CONCEPTS

- Melting
- Dissolving
- Temperature
- Change of state
- Mixtures
- Solvent
- Solute

DON'T BE SURPRISED

Melting and dissolving are two of the most often misunderstood concepts in both child and adult populations. Many children believe that when a substance dissolves in a solvent it does not exist any longer. They think it disappears. Even though they can still *taste* sugar dissolved in a drink, since they can no longer see it, it no longer exists for them. In fact, if asked if the sugar added to a liquid increases the total weight, many will say that it does not. This lack of understanding of conservation of mass is most common among elementary age students.

CONTENT BACKGROUND

Dissolving

When a substance is put into a solvent, such as water, it *dissolves* and becomes part of a homogeneous mixture, involving a *solute* and a *solvent*. The solute is the substance being dissolved and the solvent is the substance into which the solute is mixed. The result is a mixture and is a physical change that can be reversed so that the solute can be recovered again to its original crystallized form. For example, when sugar or salt are added to water, the water helps to break the solute particles down into very small molecules, which then distribute themselves equally throughout the solvent into what we call a *homogeneous* mixture. This means that if you were to sample the mixture from anywhere in the container, the ratio of sugar molecules to water molecules would be the same. Mixtures can also be *heterogeneous*. Concrete is an example of a mixture that is not homogeneous, given that gravel and sand are usually not equally dispersed throughout the mixture.

Mixtures can also involve the dissolution of gases in liquids (carbonated drinks), gases in gases (water vapor in air), and solids in solids (amalgams). In my (and probably your grandparents') day, dentists used to make fillings out of a mixture of mercury and silver. Over time, the mercury would leach out, leaving a solid hard filling of silver. As the mercury left the amalgam, there was only one place for it to go—into our bodies. However, we now know that mercury is very harmful in the human body and is no longer legally used for anything with which we may come in contact.

Various liquids dissolve solids at various rates. Temperature is a factor in that higher temperatures in the solvent almost always result in faster dissolution. If you used the story "Iced Tea" from *More Everyday Science Mysteries* (Konicek-Moran 2009), you will remember that Caroline believed that she could dissolve as much sugar in the cold iced tea as she wished. Her sister Lisa knew that this was unlikely and challenged Caroline to prove her statement. Caroline was surprised and humbled when she realized that she couldn't dissolve much sugar in the iced tea because the cold solvent was able to accommodate only a small amount of sugar before it became saturated and could dissolve no more. Temperature affects some solutes, but not all. Salt, for example, is not affected to the same extent.

Also affecting the rate of dissolution is the particle size of the solute. This is an investigable question testable using sugar cubes for example and comparing their rate of dissolving with granulated sugar, granulated sugar crushed in a mortar and pestle or confectioners sugar.

Melting

Melting is an entirely different process. Melting is a physical change that involves a transfer of heat and the changing of any substance from a solid to a liquid state. Heat from the environment is necessary to change any substance from a solid to a liquid, and then to a gas. It is very easy to see the change in state in the melting of an ice cream or ice cubes.

Water, the best known exemplar of thermodynamic change of states, can go from solid (ice) to liquid (water) to gas (water vapor) under the influence of temperature change. Some substances need extremely high or low temperatures in order to change states. In rare cases, materials such as dry ice (solid carbon dioxide) go directly to the gaseous form in a process called *sublimation*. If you have ever had occasion to watch dry ice at room temperature, you notice the white clouds appear as the solid goes to the gaseous form without going to the liquid form first. A very few metals, such as mercury, are liquid at room temperature.

So you can see that melting and dissolving are two very different things. Students who think that sucking on sugar candy or a lollipop is melting do not understand the difference between the concepts. With help, they can see that dissolving that lollipop is the same as dissolving sugar in water except that they are dissolving the sugar from the candy in saliva. The same thing happens when they chew gum. The sugar flavoring in the gum is dissolved in the saliva and swallowed.

Every child has had the opportunity to see ice cream melt and know that they have to eat quickly or lose their treat, often all over their shirts. They may find that

it is difficult to differentiate between placing ice cubes in liquid and saying that the ice cube is melting in the liquid and placing a sugar cube in water and saying it is dissolving. However, since the ice cube is a frozen liquid, the difference in temperature that occurs in the outside liquid is a direct clue that something other then dissolving is happening.

related Ideas From The National Science education standards (Nrc 1996)

K–4 Properties of Objects and Materials

- Objects have many observable properties including size, weight, shape, color, temperature, and the ability to react with other substances.
- Objects are made of one or more materials, such as paper, wood and metal. Objects can be described by the properties of the materials from which they are made, and those properties can be used to separate or sort a group of objects or materials.
- Materials can exist in different states—solid, liquid and gas. Some common materials such as water can be changed from one state to another by heating or cooling.

5–8: Properties and Changes of Properties in Matter

- A substance has characteristic properties, such as density, a boiling point, and solubility, all of which are independent of the amount of the sample. A mixture of substances often can be separated into the original substances using one or more of the characteristic properties.

related Ideas From Benchmarks For science Literacy (aaas 1993)

K–2 The Structure of Matter

- Things can be done to materials to change some of their properties, but not all materials respond the same way to what is done to them.

K–2: The Earth

- Water can be a liquid or a solid and can be made to go back and forth from one form to the other. If water is turned into ice and then the ice is allowed to melt, the amount of water is the same as it was before freezing.

3–5: *Structure of Matter*
- Heating and cooling cause changes in the properties of materials.

6–8: *Structure of Matter*
- Atoms and molecules are perpetually in motion. In liquids, the atoms or molecules have higher energy, are more loosely connected, and can slide past one another.

USING THE STORIES WITH GRADES K–4

A probe from *Uncovering Student Ideas in Science, Volume 1* (Keeley, Eberle, and Farrin 2005) called "Is It Melting?" asks students to choose from among a variety of options, those that are melting and those that are dissolving. Giving this probe to students and discussing their ideas before using the story may be a very helpful formative assessment. I cannot emphasize enough that these probes are not tests to be used to evaluate students but are assessments useful in helping you to plan your "next steps," in your teaching strategy. It helps you to see how many of your students hold the idea that the two processes are synonymous, and it may give you a glimpse of your work in trying to move them toward the current scientific view.

Asking children to participate in an ice cube melting contest is a good way to introduce the concept of melting. Give each child an ice cube in a cup and tell them to make it melt doing anything except touch it. This activity can be a type of imbedded formative assessment because watching what your students do gives you information about their understanding of the process.

Recently, I tried another tactic to see if students were aware of what happened to ice cubes when they are placed in water. Each child was given a beaker of water and then an ice cube colored green with food coloring (colored ice cubes can be made by placing a bit of food coloring in the bottom of each compartment of an ice cube tray, then adding water and mixing the solution until it is homogeneous in color). When the colored ice cube is placed in the water, the melting green water is visible as it descends to the bottom of the container (remember, cold substances sink, warm substances rise).

Given a little time, the green ice water warms a bit and begins to circulate around the beaker, mixing in. If a sample of the water is set aside and allowed to evaporate, the green color will remain behind, showing that it is a mixture—the food coloring was dissolved in the water and can still be separated. So, both melting and dissolving take place. Students can start to distinguish between the two processes by comparing them. It really is quite fun and dramatic as the dissolved food coloring drops down to the bottom of the vessel with the cool water as it melts, then swirls around the beaker as the convection currents distribute the food coloring all through the fluid after temperature equilibrium is reached.

Placing a piece of candy in a container of cool water will show that the candy dissolves and that no discernable heat was necessary to cause it to go into solution.

If you find that your children need some more information about the different properties of matter, I can recommend an article in *Science and Children* entitled, "What's the Matter With Teaching Children About Matter?" (Palmeri et al. 2008).

USING THE STORIES WITH GRADES 5–8

Students at this age also find the probe mentioned above a challenge to their thinking and their responses can provide you with valuable information about their understanding of the subject. If your students seem fairly confident about knowing the difference between melting and dissolving, it may help them to prepare an activity for younger students that makes the difference obvious. I think that we have all had the opportunity to realize that the person who is preparing the teaching activity learns more from the process than expected.

Your students may like to try to dissolve different dry ingredients such as cocoa, instant coffee, or sugar, in cold water. After that, they can try to dissolve these same ingredients in warm water.

Since a great many of your students may already have a good handle on the difference between melting and dissolving, you could try a very reliable activity that will cause them to think and rethink their conceptions—putting salt onto ice cubes. The salt dissolving in the water on top of the ice cubes (there's always a little bit of water there) will cause the cube to melt very quickly as compared to unsalted cubes. Salt dissolved in water lowers the freezing point temperature. A control should also be used so that the students can compare the two cubes. They may relate this to using salt on the highways to melt ice but perhaps have never had the opportunity to control the process and watch it unfold, firsthand. A good way to start this activity is to ask them what they think will happen if they place a teaspoon of salt on an ice cube. Have them record predictions and reasoning both in their science notebooks and publicly on the board or on a sheet of chart paper. This activity is aimed mainly at starting a discussion about both dissolving and melting, since the salt dissolves in the water but the ice cube melts. Questions may arise about the amount of salt, they may wonder if more salt will cause the cube to melt any faster. Doing this with several cubes and different amounts of salt will result in a cube "race" that will stimulate discussion.

Have them write an ending to the story with an explanation by Caroline about whether the sugar in the iced tea is dissolved or melted. Here is an embedded assessment for your use and an opportunity for the students to do some thinking about their thinking and engage in a bit of scientific literacy.

related NSTa Press Books and NSTa Journal articles

Keeley, P., 2005. *Science curriculum topic study: Bridging the gap between standards and practice.* Thousand Oaks, CA: Corwin Press.

Keeley, P., F. Eberle, and L. Farrin. 2005. *Uncovering student ideas in science, volume 1: 25 formative assessment probes.* Arlington, VA: NSTA Press.

Keeley, P., F. Eberle, and J. Tugel.2007. *Uncovering student ideas in science, volume 2: 25 more formative assessment probes.* Arlington, VA: NSTA Press.

Keeley, P., F. Eberle, and C. Dorsey. 2008. *Uncovering student ideas in science, volume 3: Another 25 formative assessment probes.* Arlington, VA: NSTA Press.

Keeley, P., and J. Tugel. 2009. *Uncovering student ideas in science, volume 4: 25 new formative assessment probes.* Arlington, VA: NSTA Press.

Konicek-Moran, R. 2008. *Everyday science mysteries: Stories for inquiry-based science teaching.* Arlington, VA: NSTA Press.

Konicek-Moran, R. 2009. *More everyday science mysteries: Stories for inquiry-based science teaching.* Arlington, VA: NSTA Press.

Konicek-Moran, R. 2010. *Even more everyday science mysteries: Stories for inquiry-based science teaching.* Arlington, VA: NSTA Press.

references

Keeley, P., F. Eberle, and L. Farrin. 2005. Is it melting? in *Uncovering student ideas in science, volume 1: 25 formative assessment probes*, 73–77. Arlington, VA: NSTA Press.

Konicek-Moran, R. 2009. Iced tea. In *More everyday science mysteries*, 169–177. Arlington, VA: NSTA Press.

Palmeri, A., A. Cole, S. DeLisle, S. Erickson, and J. Janes. 2008. What's the matter with teaching children about matter? *Science & Children* 46 (4): 20–23.

CHAPTER 15
COOLING OFF

Juan and Esteban were doing their homework at the kitchen table on a weekend afternoon. Juan was reading the problem out loud as he often did. "If you had a half cup of water in a Styrofoam cup at 25° C and added it to another half cup of water in a Styrofoam cup at the same temperature, what would be the temperature of the whole cup of water?" He paused to think a moment. "You know, Steve, it seems to be common sense that if you added the *same* temperature water to the cup, the temperature would stay the same. But that seems to be too easy."

"Yeah, you're right. That must be a trick question. You can't trust teachers not to pull a fast one on you," said Esteban.

"Maybe you add the two temperatures together so that it comes out to be 50° C."

"I guess that might work. But 50° C seems awfully hot! Somehow that doesn't sound right. I know how to *cool* off my cocoa. I just add cold milk, but I know it isn't the same temperature as the hot cocoa 'cause the milk just came out of the refrigerator."

"Right," said Juan, "and then you just have to add a little bit to cool off the cocoa."

"Now, look at the next problem. It asks us to predict the temperature of mixing two different amounts of water at the same temperature," said Esteban. "That seems like the same problem as the other one but now we have to think about the amounts."

"I can just imagine that things are going to get tougher as we go along! See. Look at the next problem. They want us to predict what the temperature would be after we mix equal amounts of water with different temperatures. I can see where this is going!"

"Well, does the amount of water make a difference?" asked Juan. "I think the temperature does, but then thinking about the amount of water messes up my brain."

"Maybe we can just do the mixing of the water and get the answers right from the thermometer."

"Hey, good idea, dude!" said Esteban. "We have everything we need to answer these questions right here in the kitchen. Then we can figure out a way to predict the answers to the other problems without having to measure everything! At least I think we ought to be able to do it. Tell you what, you use those aluminum cups over there, and I'll use these glass cups and we'll just find out by doing it. We don't have any Styrofoam cups, but what difference could that make?"

PURPOSE

The main purpose of the story is to help students understand the dynamics of heat energy transfer. Another purpose is for them to understand how to make predictions and gather and graph results from experimental data.

RELATED CONCEPTS

- Heat
- Temperature
- Thermal energy
- Thermometers
- Calorimeters
- Conduction
- Equilibrium
- Graphing

DON'T BE SURPRISED

Your students will probably not be able to distinguish between heat and temperature as concepts. It took over 2,000 years for humankind to understand the difference, but your students have your guidance and the result of many centuries of research to come to a conclusion acceptable to the current scientific community. They may also not comprehend that heat is a transfer of energy from a body that has a higher amount of thermal energy to one that has less. And it is entirely possible that even at the middle school level, students will not realize that the scientific community stresses the fact that it is heat (thermal energy) that moves and not cold (the lack of or possession of lower thermal energy).

Children have many misconceptions about heat and temperature. For one thing, children may think that heat emanates from things that warm us, like mittens and blankets. An article about this misconception (Watson and Konicek 1990) is available online at *www.exploratorium.edu/ifi/resources/workshops/teaching-forconcept.html*.

CONTENT BACKGROUND

The two boys in the story are typical students who try to solve the teacher who gave the homework rather than the problem itself! Actually, children, when presented with problems that have several numbers, often resort to adding them together (or subtracting, multiplying, or dividing them) without giving thought to the actual situation being presented. When children believe that heat and temperature are synonymous, the problems presented by the homework in the story are difficult to solve.

Heat

Children and adults alike think of heat as merely a number; the higher numbers mean hot and the lower ones, cold. Heat and cold are separate entities that fight it out in a substance until one wins. We often add to this confusion by telling them to close doors to "keep the cold out." Sloppy use of language conflicts with the language of physics. We are dealing with the laws of thermodynamics here in this story and as complex as that sounds, the reasoning is quite simple. Heat is defined as the transfer of kinetic energy from a warmer body to a cooler one. Heat is often called *thermal energy* and is a property of all matter since there are always moving molecules in matter. While physicists tell us that absolute zero is reached when no more heat can be extracted from matter, they have not been able to reach that point in the laboratory. They've come close, but there's been no cigar.

We've all been raised on the statement "energy can be neither created nor destroyed," the first law of thermodynamics, and this makes sense. Energy can, however, be converted from one form to another. And it can be moved from one place to another: through conduction (touching), radiation (from radiating bodies like the sun or an electric heater), and convection (through currents that move it around).

In the case of conduction, kinetic energy from moving atoms and molecules in one substance pass on their energy to other atoms and molecules in another substance by bouncing into them and giving them a "kick," transferring energy as they do. This transfer will continue until both giver and receiver have reached equilibrium. In fact, it never really stops because the amount of heat exchanged will continue since the temperatures of the bodies involved are constantly changing.

When you stand in the sun or near a stove that is giving off heat radiation, you can feel the infrared energy waves reaching you and warming you. All bodies radiate heat to some extent and do not need air as a medium to transfer it. We have evidence of this since the Sun, our source of radiational heat, sends its thermal energy to Earth through almost completely empty space.

To describe convection, it helps to consider an example. If you have a pot of water on the stove and turn on the burner beneath it, the pan will absorb heat from the burner and transfer it by conduction to the water at the bottom of the pan. As the water warms, it becomes less dense than the rest of the water and rises to the top of the pan. The cooler water at the top sinks down to take its place, and is in turn warmed so that it too rises. Meanwhile, the water that rose previously has cooled somewhat, so it sinks and then warms, and so on. A circular current is set up until all of the water shares an equal amount of thermal energy. The same thing can happen in a room, since physicists consider air to behave like a fluid. Convection does not occur in solids.

Temperature

Temperature measures changes in heat. It is a completely arbitrary number that depends entirely on which scale one is using. There are three main temperature scales in use today, the Fahrenheit, Celsius and Kelvin scales. Each is named for its originator and thus is capitalized:

- Fahrenheit for Gabriel Fahrenheit in 1719 (boiling point of water: 212°F, freezing point: 32°F)
- Celsius for Anders Celsius in 1742 (boiling point of water: 100°C, freezing point: 0°C)
- Kelvin for Lord Kelvin of Scotland in 1848, (absolute zero is 0 K, (-273° C), boiling point of water: 373°C, freezing point of water: 273°C)

Note: *Absolute zero* is defined as the temperature at which no heat can be extracted, or in some circles, the point at which all molecular motion ceases. As we noted before, this temperature has never been reached even in controlled laboratory settings.

Thermometers only became useful for measurement of changes in temperature when Santorio Santorio (1611) added a scale to an instrument called a thermoscope, invented by Galileo in 1596. This had only given indications of a rise or fall in temperature due to the expansion of the substance in which it was immersed. During the following centuries, other instruments to measure heat called *thermometers* were developed, but unless scientists adopted some sort of an agreed upon *scale* or *calibration* the measurements from these were useless. In 1724, Fahrenheit developed the scale that is still in use today in a very few countries, based on events that were familiar to people and that could frame reference points (like the freezing and boiling points of water). In science, the Celsius scale, which is based on the freezing point of water being 0° and boiling point 100°, thus making calculation easier, is considered the standard scale and was adopted as such in 1948. Lord Kelvin's scale was formed to take absolute zero into account. So the thermometer and the measurement of temperature was less an invention than a series of developments and improvements over the years. They have evolved.

When a thermometer is placed in a liquid or under the tongue, it is really registering its *own* temperature since it has come in contact with molecules that are transferring kinetic energy. In essence, it is a snapshot of the kinetic energy in any given substance in a specific part of the container that holds the substance. It does *not* measure the *total* energy of the substance. An example of what this means follows. A cup of boiling water (100° C) and a single drop from that cup both have the same temperature. I would be willing to place one single drop of that water on my hand without fear of scalding myself. I would not be willing to pour the entire cup of boiling water on my hand, even though both the drop and the cup of water have the same temperature. The amount of heat in the cup of water far exceeds the amount of heat in the drop of water because there is more of it. The heat from the entire cup of water would do serious damage to my hand whereas the single drop would merely cause a pinprick of feeling. This is why measurement of heat is better done with the concept of *calories* or *joules*, which I will explain below. (CAUTION: Do not try this with children! Use it as an example but caution them not to play with boiling liquids in any form. Accidents can happen!)

Oftentimes children, and even adults, confuse heat and temperature because when the heat or kinetic energy in a substance changes, the temperature changes. But temperature measures just kinetic energy, while heat measures the total energy of a substance, including potential energy (more about this below). It was not until Joseph Black, a Scottish scientist explained the difference between heat and temperature in 1761 that science could finally move forward. When scientists thought that the two were synonymous, they were at a loss to explain many phenomena. The discoveries and theories of Joseph Black led the way to progress in the area of thermodynamics.

Measuring Heat

First, let me say again, that all matter is made up of constantly jostling molecules and atoms. Their motion means that there is kinetic energy present in all matter (the energy of motion is known as *kinetic energy*). You cannot see this motion any more than you can see the molecules themselves, but, according to the present theory, all matter is made up of these moving particles, a sort of hidden energy. The more motion, the more heat.

What we refer to loosely as heat is actually *thermal energy*. Thermal energy is the sum total of the energy in any given substance, including the *kinetic* energy in the molecules and atoms and the *potential* energy in the substance (although this can get very complicated and we won't get too deep here). It is very common to refer to a substance as containing heat, but this is incorrect. Matter does not contain heat but it does have moving molecules: molecular kinetic energy and the potential for more molecular movement if necessary. Heat flows from one thing to another and it is that kinetic energy that is being transferred. This may seem picky but it is an important distinction in thermodynamics. When there is contact between two objects, heat always flows from the substance that has more kinetic energy to the substance that has less. In other words, heat flows from warmer matter to cooler matter. It continues to do so until both bits of matter are at the same temperature. It is then that we can say that they have reached equilibrium.

Thermal energy is measured in an agreed upon unit called a *calorie*. One calorie is defined as the amount of heat required to raise the temperature of one gram of water to one degree Celsius. If you see the Calorie spelled with a capital C, it is a kilocalorie and is the amount of heat required to raise the temperature of one kilogram of water, one degree Celsius. You may have also seen heat energy measured in *British Thermal Units* or *btu's*. This is the amount of thermal energy needed to heat 1 pound of water to 1 degree F. (Interesting note: The *joule* is now considered to be the international unit of thermal energy, except in the United States. A joule is the force of one Newton moving an object one meter, but this translates into energy as well. Once again, we won't go too deep here—it's just important that you get an idea of the current terminology.)

CHAPTER 15

You may be familiar with the calorie in measuring the food that you eat. Any food has the potential of releasing heat, particularly when it combines with oxygen, either in a flame or in your blood stream. Various kinds of food have different caloric values, which are determined by burning them in a *calorimeter*, an instrument that determines how many grams of water can be raised one degree Celsius, or more important to us, put unwanted pounds on our bodily frames.

We all know that when we touch objects in a room that some feel warmer or cooler than others. If they have been in the same environment for a long time, they have all come to equilibrium and are the same temperature. Yet, when we touch something, metal for instance, it feels cooler than wood or cloth. Some materials transfer heat more readily than others. Some accept heat more readily than others and by the same token give it up more easily. When we touch an object that transfers heat easily, heat from our bodies flows into that object and our hands feel cooler than when we touch something that does not transfer heat easily. In other words, our hands are not good thermometers—they are easily fooled. By the same token, you cannot feel your own forehead to see if you have a fever since your hand and head are probably the same temperature and no heat flows. If your hands are cold, heat from your forehead will flow to your hand but this is not an indication of the true temperature of your body.

Another example of this difference in heating and cooling capacity is found in your refrigerator. Juices in aluminum cans cool much faster than those in glass containers. Physicists say that aluminum has a lower *specific heat* than glass (*specific heat* is—loosely—a measure of the capacity of substances to hold heat). The aluminum cans allow heat to leave the juice easily while the glass containers don't allow heat to leave quite so fast. However, the reverse is true as well. Once out of the refrigerator, the aluminum cans of juice will absorb heat much faster and have a warmer temperature sooner than glass.

Water has a very high specific heat because it takes a great deal of heat energy to raise its temperature. It loses its heat slowly, which makes it a perfect medium for hot water bottles because the water in the rubber container keeps feet warm for a longer time.

Either way, the laws of thermodynamics are part of your everyday life. Thermal energy and the way that it moves is important in many ways. It is best that we understand it and use our knowledge to save money and energy.

related ideas from the national science education standards (NrC 1996)

K–4: *Properties of Objects and Materials*

- Objects have many observable properties, including size, weight, shape, color, temperature and the ability to react with other substances. Those properties can be measured using tools, such as rulers, balances, and thermometers.
- Objects are made of one or more materials, such as paper, wood and metal.
- Objects can be described by the properties of the materials from which they are made, and those properties can be used to separate or sort a group of objects or materials.

K–4: *Light, Heat, Electricity, and Magnetism*

- Heat can be produced in many ways such as burning, rubbing and mixing one substance with another. Heat can move from one object to another.

5–8: *Transfer of Energy*

- Energy is a property of many substances and is associated with heat, light, electricity, mechanical motion, sound, nuclear energy, and nature of a chemical change. Energy is transferred in many ways.
- Heat moves in predictable ways, flowing from warmer objects to cooler ones until both reach the same temperature.

related ideas from benchmarks for science literacy (aaas 1993)

K–2: *Energy Transformation*

- The Sun warms the land, air, and water.

3–5: *Energy Transformation*

- When warmer things are put with cooler ones, the warm ones lose heat and the cool ones gain it until they are all the same temperature. A warmer object can warm a cooler one by contact or at a distance.
- Some materials conduct heat much better than others. Poor conductors can reduce heat loss.

6–8: Energy Transformation

- Heat can be transferred through materials by the collision of atoms or across space by radiation. If the material is fluid, currents will be set up in it that aid the transfer of heat.
- Energy appears in different forms. Heat energy is in the disorderly motion of molecules.

USING THE STORIES WITH GRADES K–4

I prefer to find out what my students already think about mixing things of different temperatures and usually give the probe, "Mixing Water," from *Uncovering Student Ideas in Science, Volume 2* (Keeley, Eberle, and Tugel 2007). This probe asks students to predict the resulting temperature when two tumblers containing the same amount of water at different temperatures are mixed. This will tell you if your students are merely manipulating numbers or are aware that the resulting temperature will be somewhere between the high and low temperatures of the two tumblers.

For younger students, a modified probe might ask if the temperature of the mixture will be different than either of the originals. After trying it out with temperatures that are not dangerous to the children's safety, such as room temperature with cool water, ask them how they think the resulting temperature happened. You can introduce them to the word *heat* and talk about its being transferred from the water in one glass to the water in the other. Since the resulting mixture was warmer than the colder water you might try to elicit from them that the heat traveled from the warmer to the cooler water.

By asking them to list the things that they know heat up and cool down, you can make a list of objects with which they have had direct experience. Our experience is that the list will look something like this:

- Water heats up when it is in the sunshine.
- Sand heats up when it is in the sunshine.
- Our house heats up when it is in the sunshine.
- Metal things heat up when they are in the sunshine.
- Things cool off when they are in the refrigerator.
- Some things cool off faster or heat up faster than others.

Taking this list and turning the statements into questions that are investigable is a simple task, e.g., Does water heat up when it is in the sunshine? Do some things cool off or heat up faster than others? This sets the stage for investigations, predictions and conclusions. Read the *Science and Children* article "Heating Up, Cooling Down" (Damonte 2005) for more ideas on how to involve students in the mystery of how different materials take in heat energy and transfer it to other things. The author looks at common materials like sand and water to make suggestions about inquiry into heat and especially specific heat differences.

Older elementary students, because they can read and interpret thermometers can quantify the changes in such investigations. Perhaps you have digital thermometer probes that you can use to recover data on temperature change over time. If not, simple thermometers with data recorded in science notebooks will generate wonderful discussions. You can also take the opportunity to introduce them to graphing their results. More will be said about that in the next section for middle school. You may even want to try the calorimeter activity described below with older children.

USING THE STORIES WITH GRADES 5–8

Having recently spent a day with a group of talented middle school teachers from Springfield, Massachusetts, investigating the topic of heat and heat transfer, I am prepared to discuss some of the activities and ideas that excited them and which they plan to use in their inquiry activities in their classes. Middle school children have skills that allow them to collect data and graph the results. However, sometimes it is necessary for the teacher to help them interpret graphs. This "teachable moment" often occurs when they have plotted their data and are confused by the result.

After the students have had a chance to try a few different scenarios like those listed in the story, you can ask them if they have any ideas about the direction in which heat moves. This simple calorimeter activity can measure heat transfer from one insulated container to another. The diagram (Figure 15-1) shows how the setup is arranged. Materials for this activity are:

- Two insulated (Styrofoam) cups with foam tops made from scrap foam material. The foam should have two slits, one for a thermometer and the other for the aluminum bar described below.
- One aluminum "bar" made from a folded strip of aluminum foil long enough for the ends of the bar to be immersed in water in both cups.
- Two thermometers long enough to reach the water in each container and also long enough to be read without removal from the container.
- Two cups of 100 cc of water at temperatures approximating 60°C and 30° C, 100 cc of each will be placed in the two separate cups.
- One chart for recording the temperatures of each thermometer at two-minute intervals.
- A copy of the probe for each student.
- A copy of the setup diagram for each group of students.

Figure 15.1 This is the set-up for the calorimeter activity. The bar that is embedded in each Styrofoam cup is made from folded aluminum foil wrap. The cups have foam or other insulation on top and the bar and thermometers are inserted through slits in the foam tops.

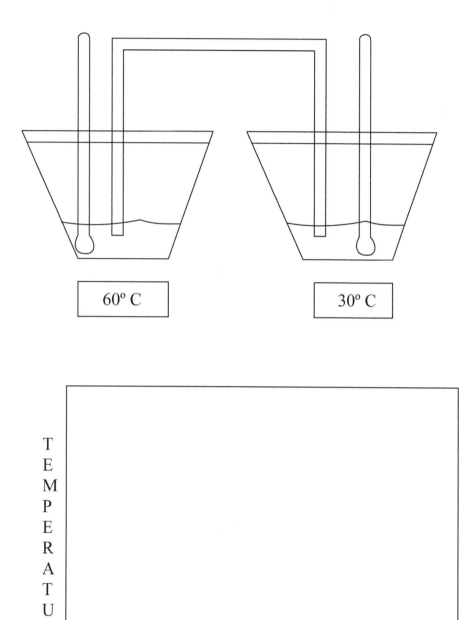

60° C

30° C

TEMPERATURE

TIME

You will need to heat some water to approximately 60° C for one cup and use room temperature water of approximately 25–30°C for the other. I suggest that children work in groups of three or four. When you are ready to begin, have the students pour the water into the two containers as shown, then put the cover, the aluminum bars and thermometers into place. Have the students take an initial temperature reading in each cup; then record that and the temperatures in each cup at two-minute intervals, keeping the data on a data sheet for use later when they graph their results.

Figure 15.2 Four students were using two Styrofoam cup calorimeters to find out how heat is conducted from one calorimeter to another. Each cup had water at a different temperature. In one styrofoam cup, there were 100 cc of wate at 90°C and in the other, the same amount of water at 30°C. An aluminum bar connected the two cups and the water they held. A thermometer was inserted in each cup to record the temperature. Readings of the temperatures were recorded every two minutes and were plotted on a graph, time against temperature. Below are four pictures of possible graphs after 30 minutes of recording. Which of the four graphs do you think most depicts the one the students created after plotting their data? Circle the letter you choose and explain your choice.

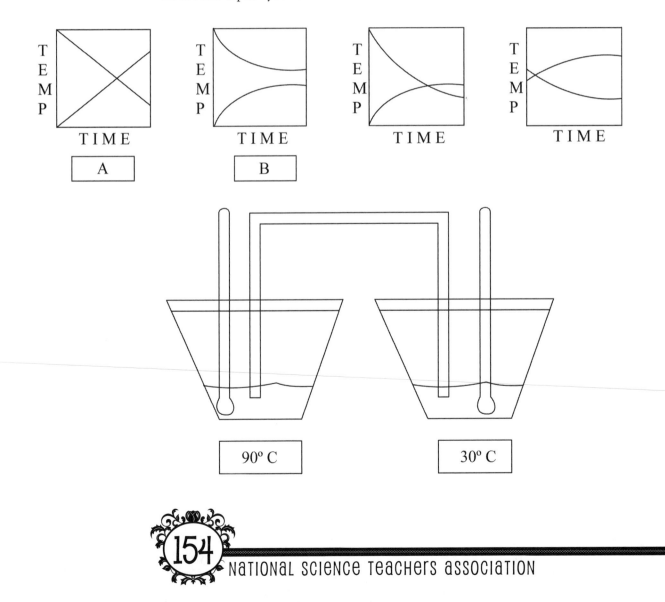

But, before the activity explain to them what they are going to be doing, observing, and finally graphing. Tell them that they are going to see if the energy in one cup influences the temperature of the water in the other cup via the bar that connects them. Then, before the students begin the activity, ask them to take the probe I have written. (See Figure 15.2) It asks them to predict what a graph of their findings will look like. Ask them to discuss their predictions and give reasons for their choices.

Those who understand what is about to happen will lean toward choice (B) because they will predict that the cooler water will be warmed by the heat traveling from the warmer cup via the aluminum bridge. They may still be confused as to why the lines in (B) never meet but once they perform the activity they will probably touch the aluminum bridge and feel the heat. This is heat that is lost to the atmosphere and so does not make it into the cooler cup. This means that all of the heat energy in the warm cup does not make it into the cooler cup and the actual temperature at equilibrium will be somewhere in between the temperatures shown as the lines in the graph level off and do not change. It would be somewhere around 45°C if no heat were lost. It also shows that it is the heat that is moving since the bridge gets warm and not cool. The class discussion here is very important since it gives those with misconceptions the chance to voice their ideas and to listen to others. I have found that students are more likely to listen to expressed ideas and modify their own if they are from a peer rather than a teacher.

You will notice that the correct graph is the one where the lines never meet but reach equilibrium before the lines touch. It is (B) on the probe. The others are not correct because in (A) and (C) they show the lines crossing which could never happen due to the impending equilibrium or (D) they start at the same place and both lose and gain heat at the same rate.

This activity improves both the conceptual understanding of heat transfer and students' ability to produce and interpret graphs. They can then finish the story by having the boys come to a conclusion about mixing water. I predict that some student will suggest mixing the two cups of water to see what the resulting temperature is without the aluminum bar setup.

If you would like to continue this investigation, challenge the students to create a setup that will cut down the heat loss due to the inefficiency of the system they just used. This will lead them into the area of finding materials that reduce heat loss, insulators. They will then likely come closer to achieving a graph that is more like ideal conditions.

For a look at how one teacher combined activities on heating and cooling and introduced the idea of specific heat as insulators, see the *Science Scope* article "To Heat or Not to Heat" (May and Kurbin 2003) in the journal archives at *www. nsta.org*.

ReLaTeD NSTa PRESS BOOKS AND NSTa JOURNaL aRTICLES

Childs, G. 2007. A solar energy cycle. *Science and Children* (Mar.): 26–29.

Konicek-Moran, R.2008. Everyday science mysteries. Arlington, VA: NSTA Press.

Konicek-Moran, R. 2009 More everyday science mysteries. Arlington, VA: NSTA Press

ReFeRences

American Association for the Advancement of Science (AAAS).1993. *Benchmarks for science literacy*. New York: Oxford University Press.

Damonte, K. 2005. Heating up, cooling down. *Science and Children* 42 (9): 47–48.

Keeley, P., F. Eberle, and J. Tugel. 2007. Mixing water. In *Uncovering student ideas in science, volume 2: 25 more formative assessment probes*, 83–89. Arlington, VA: NSTA Press.

May, K., and M. Kurbin. 2003. To heat or not to heat. *Science Scope* 26 (5): 38–41.

National Research Council (NRC). 1996. *National science education standards*. Washington, DC: National Academy Press.

Watson, B., and R. Konicek. 1990. Teaching for conceptual change: Confronting children's experience. *Phi Delta Kappan* 71 (9): 680–685.

CHAPTER 16
PARTY MELTDOWN

K elsey and her friends were having a party—
a big party! There were to be 12 girls, lots
of noise and music, pizza, chips, cake, and
soda. Fortunately there would be plenty of
ice since Kelsey and Lucy had been mak-
ing ice all day in preparation for the drinks. They had
borrowed ice cube trays from everyone in the neighborhood and filled the freezer with the trays full of water.

By the time the girls began to arrive, the freezer had
done its job and there was plenty of ice. They then put
the ice cubes into any kind of bowls they could find.
Some were metal and others were plastic or glass. They
all went into the freezer to keep cold.

Kelsey was in the kitchen getting more bowls of
ice out of the freezer when Elizabeth popped her new
Justin Beiber CD into the player. As soon as he was

recognized, everyone screamed with excitement and ran into the family room to dance. This, of course, included Kelsey, who in the rush left two bowls of cubes out on the kitchen counter.

Much later, when it was time for more soda and more ice, the girls returned to the kitchen to find the two bowls sitting there where they had been forgotten.

"Hey, one bowl is full of ice cold water but no ice," Nicole exclaimed, "but the other one still has some ice in it. Bet you forgot to put the metal one in the freezer."

"No way," Kelsey responded. "We took them out at the same time and forgot them when we went in to dance!"

"Puh-lease!" said Nicole. "That's why one is water and the other is still half frozen. Get real, Kelse!"

"Oh, forget it!" exclaimed Kelsey. "Just grab whatever ice is left and let's get back to the party."

The next day, as Kelsey, Lucy, and Sara were cleaning up after the party, the two bowls were still on the counter and Kelsey reminded Lucy of what had happened the night before.

"Duh," said Lucy. "The ice cubes in the plastic bowl get colder in the freezer than the cubes in the metal tray and since they were colder, they melted slower."

"That's crazy, the ice cubes were all the same temperature," said Sara, who often had strong opinions. "The metal bowls are shiny and shiny stuff gets hotter than dull stuff like plastic. So the shiny bowl got hotter and melted the ice. All of your cooking pans are shiny. Think that's just luck?"

"Metal things just naturally get hotter quicker," said Kelsey. "It has nothing to do with shiny or dull! Don't you remember how hot metal things get out in the sun? The metal bowl took up heat from the room air and melted the ice."

"Who cares!" laughed Lucy. "Let's get the dishes washed and get out of this kitchen!"

They did just that, but inside Kelsey's head, the question remained. Who was right? Or were any of them? Did any theory make more sense than another?

PURPOSE

This story will allow students to become aware that some materials give up and take in thermal energy at different rates. A second purpose is to understand how two different substances having the same temperature can feel as though they have different temperatures because of the various rates of the movement of thermal energy between the hand and the substance.

related CONCEPTS

- Heat
- Thermal conductivity
- Thermal energy
- Temperature
- Conduction
- Change of state (melting)

DON'T BE SURPRISED

Students often focus more on the visible properties of objects, like shininess, than the material of which they are actually made. Many will side with Lucy and hold conflicting opinions that even though the ice cubes were in the freezer together, the cubes in the plastic bowl got colder than the ones in the metal bowl. It is not uncommon for students to hold conflicting opinions without realizing it.

CONTENT BACKGROUND

Heat is transferred as energy (thermal energy) from warmer objects to cooler objects by one of three different processes: conduction, convection, or radiation. In this story we are dealing with the process of heat transfer by conduction. Conduction involves the molecules or atoms of one substance coming in contact with their counterparts in another substance. In this instance, the girls argued about the ice cubes that were touching the metal and plastic bowls. An additional conduction of heat took place when the surface of the bowls contacted room temperature air.

Not all substances conduct thermal energy equally. Some materials such as metals conduct energy more efficiently than nonmetals. They are said to have a greater *thermal conductivity*. Thermal energy by conduction involves touching. The molecules of the warmer substance are moving faster with more energy than the molecules of the cooler substance. When the molecules from the warmer substance collide with the molecules of the cooler substance, some of that energy is transferred to the slower molecules causing them to move faster. They, in turn, interact with their cooler neighbor molecules and transfer some of their energy along until the molecules all have pretty much the same amount of energy. We

can gauge with a thermometer or sense with our bodies the average of this thermal activity—what we call an object's *temperature*. Thus, temperature is a measure of the average amount of energy in a system but is not a measure of the total amount of energy in that system.

This form of energy transfer is the basis of warming and cooling used in everyday life. Heat energy moves from the warmer to the cooler until both substances or areas are in equilibrium (or are the same temperature). You will notice that I did not say that they have the same amount of energy because that will depend upon the mass of either of the systems involved. A swimming pool at room temperature will have much more heat in it than a cup of boiling tea. Their temperatures are very different, but the swimming pool holds a great deal more energy due to its greater mass.

Another way to look at this relationship to mass and temperature is to think of a cup of boiling tea and all of the energy in that cup. You would probably not mind having a drop of that tea land on your hand but you certainly would not want the entire cup. They are both the same temperature, yet the cup has much more energy and could do much more damage to your hand if the whole cup were to be spilled on it. A drop might sting for a moment but would not do any damage.

In the story, the girls placed the ice cubes in separate bowls in the freezer which we can presume had a constant temperature. If the bowls were in the freezer for a reasonable amount of time, everything would then have the same temperature, freezer air, ice, and bowls alike. They were removed from the freezer to the counter and subjected to room temperature air. Since thermal energy moves from the warmer to cooler, we can assume that the energy in the room air, by conduction, transferred to the bowls and then again by conduction to the ice cubes.

Now, this is where the differences in thermal conductivity come in. The metal bowl has a greater thermal conductivity than the plastic, meaning the molecules of the metal transfer thermal energy faster than the plastic. Therefore, the thermal energy reaches equilibrium with the ice in the metal bowl before equilibrium is reached in the plastic bowl. So, the plastic bowl still had ice in it, while the metal bowl had water from melted ice.

One of the most puzzling thing in everyday life is that when we touch a metal object in a room, it feels colder than a wooden or plastic object in the same room. Both objects have been in the room for some time and and therefore should be the same temperature. Yet, the metal feels colder. Think about this before you read on.

Heat flows from our body to the metal and wood because of the difference in temperature. We are warmer than room temperature (unless we happen to be in a sauna). Heat leaves our body at a faster rate when in contact with the metal than it does when in contact with the wood. Thus, it feels like the metal is cooler. So, we are fooled into believing that the metal is cooler.

It is interesting to note that the greater the temperature difference between two substances, the faster the temperature change toward equilibrium. This temperature difference is often referred to as ΔT or *delta T*. Your cup of tea will cool

off faster when it is at a higher temperature than if it is at a lower temperature. This makes perfectly logical sense since the higher energy molecules bouncing around there are, the more likely they will leave the surface of the tea and take heat away from the drink.

When I was a kid, I grew up during the time, before the Salk vaccine, when parents' worst fear for their children was the disease polio. Every time there was a suspicion of a fever, parents worried. Countless times, my parents put their foreheads against mine to see if I had a fever. If my forehead was warmer than theirs, the heat from my body would be very evident to their cooler bodies. The same principal causes thermometers to work in that they absorb heat from or give up heat to the substance into which they are placed. With exposure to heat, the liquid in the thermometer will expand through conduction and with exposure to cooler substances, the liquid inside will contract as it gives up heat. So you see, heat transfer via conduction is everywhere in our world and can explain so many everyday mysteries.

related ideas from the National Science education standards (NRC 1996)

K–4: Properties of Objects and Materials

- Objects have many observable properties, including size, weight, shape, color, temperature and the ability to react with other substances. Those properties can be measured using tools, such as rulers, balances, and thermometers.
- Objects are made of one or more materials, such as paper, wood and metal.
- Objects can be described by the properties of the materials from which they are made, and those properties can be used to separate or sort a group of objects or materials.

K–4: Light, Heat, Electricity, and Magnetism

- Heat can be produced in many ways such as burning, rubbing and mixing one substance with another. Heat can move from one object to another.

5–8: Transfer of Energy

- Energy is a property of many substances and is associated with heat, light, electricity, mechanical motion, sound, nuclear energy, and nature of a chemical change. Energy is transferred in many ways.
- Heat moves in predictable ways, flowing from warmer objects to cooler ones until both reach the same temperature.

related ideas from benchmarks for science literacy (aaas 1993)

K–2: *Energy Transformation*
- The Sun warms the land, air, and water.

3–5: *Energy Transformation*
- When warmer things are put with cooler ones, the warm ones lose heat and the cool ones gain it until they are all the same temperature. A warmer object can warm a cooler one by contact or at a distance.
- Some materials conduct heat much better than others. Poor conductors can reduce heat loss.

6–8: *Energy Transformation*
- Heat can be transferred through materials by the collision of atoms or across space by radiation. If the material is fluid, currents will be set up in it that aid the transfer of heat.
- Energy appears in different forms. Heat energy is in the disorderly motion of molecules.

USING THE STORIES WITH GRADES K–4

Very young children will be able to understand the changes of state in substances like water when it goes from liquid to ice or liquid to gas, or vice versa. I like to give young children an ice cube in a plastic tray and ask them to engage in an ice cube race to see who can melt the ice cube fastest without touching it. Witnessing the change of state is paramount for understanding this concept when they are older. Children of early grades never seem to tire of watching state changes and can be introduced to the idea of heat energy coming in to play in these changes.

In this case, you might ask children to choose from several different kinds of trays made of different materials on which to place their ice cubes. This imbedded formative assessment will tell you which children are aware that certain kinds of trays that are better conductors of heat and will make a difference in how fast their ice cube melts. If they choose a good conductor of heat, they will have a head start on melting their ice cube regardless of what else they do to speed up the melting.

NSTA members can access online the article "Heating Up, Cooling Down" (Damonte 2005) from the NSTA website. This article looks at the difference in *specific heat capacity* of three substances: sand, water, and soil. Specific heat capacity is the amount of heat needed to raise the temperature of a substance. Every substance has its own specific heat capacity. A given amount of water, for example,

requires five times more heat to raise its temperature than the same amount of sand, and by the same token, gives off heat five times more slowly. You may have noticed this if you have walked across a sandy beach on a hot day. The heat transfer to your feet was instantaneous and painful because the sand heated up more quickly than the water where you probably sought relief!

Upper elementary students can follow the story as it develops and duplicate the situation. This will convince them that the problem is plausible. Then they can test containers made of different substances to see the variation in the rate of melting. Of course, you must help them list the results, and keep all of the variables controlled.

For a quick and dramatic introduction to the issue of heat transfer, you can place an ice cube on an inverted metal can, and another on an inverted plastic, or better yet, Styrofoam cup of the same size (it helps if you remove any paper label on the can so that more surface area can be exposed to the air). The can will pick up the heat from the air in the room much more quickly than will the cup, and the ice cube on the can will begin to melt almost immediately while the other cube will remain in its frozen state for much longer. The discussion that follows should bring out the fact that the ice cubes were placed on two different materials. You could ask, "What do you think would happen if you wrapped your hands around both cups?" This is the time to introduce the terminology of heat energy and transfer (conduction), to the students because the vocabulary will be in context with the situation. Your students should now be able to finish the story of the mystery of the two bowls of ice cubes.

USING THE STORIES WITH GRADES 5-8

Middle school students may also want to duplicate the story to verify the outcome. However, a good discussion should precede the repeat of the story so that you can assess their conceptions of heat and heat transfer. Listening to your students as they talk is a wonderful way to understand their thinking, and it can guide you toward the next steps. Some of the ideas listed in the section for K–4 may be useful, and the article by Damonte (cited on p. 160) would also be of help to you in attempting to move the students toward a better undertanding of the concepts involved. Since students at this age are familiar with thermometers, the difference between heat energy and temperature will be even more accessible to them. The drop vs. the cup of boiling water I talked about in the Content Background section will help to make the distinction more real.

Using the can and Styrofoam cup activity mentioned above can also be a discussion stimulator. However, the story is an investigable activity in itself and will give the results that allow the students to finish the story. They can try the various suggestions made by the girls about shiny, dull (plastic), metal, and other holders and the idea of how cold the ice can get if it is placed in the freezer. If you place a thermometer in the water before you freeze the ice in the ice cube tray, it can give you the internal temperature of the ice cubes. Students will find that the ice cubes will all be the same temperature as the surrounding air in the freezer regardless of the container. This negates Sara's and Lucy's claims, but Kelsey's idea should prove to be true.

A very interesting extension using ideas about heat and heat transfer is described in the *Science Scope* article "Science Sampler: A Slice of Solar" (Galus 2003). The author asked her students to design solar ovens to cook an apple slice using their recently acquired knowledge of heat transfer while using common everyday materials. This kind of activity helps students to solidify their knowledge through application.

In 2010, there was an International Boiling Point Project in which students from all over the world entered data on what factors they found influenced the boiling point of water. Although the project may not continue further, the website holds a great deal of information which might be useful to you in teaching about heat, heat transfer, and state change. The website is *www.ciese.org/curriculum/boilproj/*.

related NSTa Press Books and NSTa JouRNaL aRTicLes

Keeley, P. 2005. *Science curriculum topic study: Bridging the gap between standards and practice.* Thousand Oaks, CA: Corwin Press.

Keeley, P., F. Eberle, and L. Farrin. 2005. *Uncovering student ideas in science, volume 1: 25 formative assessment probes.* Arlington, VA: NSTA Press.

Keeley, P., F. Eberle, and J. Tugel.2007. *Uncovering student ideas in science, volume 2: 25 more formative assessment probes.* Arlington, VA: NSTA Press.

Keeley, P., F. Eberle, and C. Dorsey. 2008. *Uncovering student ideas in science, volume 3: Another 25 formative assessment probes.* Arlington, VA: NSTA Press.

Keeley, P., and J. Tugel. 2009. *Uncovering student ideas in science, volume 4: 25 new formative assessment probes.* Arlington, VA: NSTA Press.

Konicek-Moran, R. 2008. *Everyday science mysteries: Stories for inquiry-based science teaching.* Arlington, VA: NSTA Press.

Konicek-Moran, R. 2009. *More everyday science mysteries: Stories for inquiry-based science teaching.* Arlington, VA: NSTA Press.

Konicek-Moran, R. 2010. *Even more everyday science mysteries: Stories for inquiry-based science teaching.* Arlington, VA: NSTA Press.

May, K., and M. Kurbin. 2003. To heat or not to heat. *Science Scope* 26 (5): 38–41.

references

Damonte, K. 2005. Heating up, cooling down. *Science and Children* 42 (9): 47–48.

Galus, P. 2003. Science Sampler: A slice of solar. *Science Scope* 26 (8): 56–57.

International Boiling Point Project. 2009. *www.ciese.org/curriculum/boilproj/*

THE CROOKED SWING

Allan and Serena were walking home from school when they came upon a new garden swing in their next-door neighbor's yard. The swing was attached to the limb of a large tree. (Look closely at the picture on page 163 so that you can see the way the swing was hung.) The two kids could not resist trying it out so they asked their neighbors, an older couple named Mr. and Ms. King, who were out doing yard work, if they minded. Ms. King answered their request with a kindly smile.

"Certainly you may use it. But I think you should know that we've found a problem with it. You'll see when you try it. Maybe you can help us to solve the problem."

"Thank you. We'll do our best."

Allan sat on the swing on one side, and drew it back and let it swing for a while.

"Hey, I see the problem," he said. "It swings crooked, not back and forth like a normal swing. Maybe if you got on with me, Serena, the extra weight will help."

Serena got on the swing next to him and the result was the same. The swing would just not swing straight.

"Hey," she said, "I wonder if it has anything to do with the swing being a kind of pendulum? We've been studying them in school. Maybe there is something wrong with it."

"Well, there's a way to make the swing higher off the ground. We can shorten the chains and see if that will help," said Allan.

"Okay," said Serena, and they tried to shorten the swing by raising it a few links of chain on both sides.

"Now let's try it," said Allan.

They did and got the same results.

"Rats," said Serena," I thought that might do it. I guess I need to learn a little more about pendulums. I thought I knew enough, but this is different in some way. Maybe we ought to try to make a model of this swing and see what makes it go crooked."

In a few days, they again knocked on their neighbor's door and said, "We think we know what your problem is with the swing. And we think we can help you fix it."

PURPOSE

This story has two purposes. One is to apply what is known about pendulums to a new problem and the other is to use technical skills to solve a problem. The swing operates on the principles of periodic motion and the crooked branch, upon which the swing is hung, presents a technical problem looking for a solution.

RELATED CONCEPTS

- Periodic motion
- Problem solving
- Model building
- Analysis of problems
- Creativity
- Systems

DON'T BE SURPRISED

Students may not see the relationship between what they know about pendulums and the problem in the way the swing is hung from the crooked branch. People often have difficulty seeing a problem as a connection of parts. The swing is a system and the relationship between the parts of the system is relevant to solving the problem.

CONTENT BACKGROUND

Students need to know something about pendulums in order to see the problem of the crooked swing. The swing is a *system* made up of the chair, the chains that support it and the tree limb from which the chains are suspended. This is, in essence, a coupled pendulum, each end of the swing being a simple pendulum and responding to the laws of *periodic motion*. Periodic motion is any motion that repeats itself and has an identical time interval called a *period*. The period is defined as the time it takes for a regularly occurring motion to complete one cycle. This definition includes the motion of the planets around the Sun, the movement of shadows throughout a day and the pendulum in a grandfather's clock. It could also include waves in water, a bouncing ball, or the vibrations of a tuning fork. Obviously it also includes a swing, as in this case, and it is the periodic motion of the two ends of the swing that must be synchronized in order for them to move together. Another way to put it is that the periods of the two ends of the swing must be equal for the swing to move in a straight path. If the period of one pendulum on one end of the swing is not the same as the other, one end will swing faster/slower than the other, resulting in a crooked path.

Here is a quick review about pendulums. If you have a copy of *Everyday Science Mysteries* (Konicek-Moran 2008), there is a chapter and story called "Grandfather's

Clock" in which a girl tries to find out how to get her new clock to keep time correctly. This story and teacher material can be of great value to you in preparing to teach this concept. A *pendulum* is a system consisting of a point where a rod or string is attached; at the end of the rod or string is a weight called a *bob*. Thus, the system swings back and forth when energy is applied to the pendulum so that it swings in a consistent arc. The variables that can be changed are: the weight of the bob, the length of the pendulum from swing point to the center of the mass of the bob and the height of the release point of the swing (called *amplitude*). The vital questions about pendulums are: does the weight of the bob, the size of the arc of the swing, or the length of the system have any effect upon the period of the pendulum? It turns out that only the length of the pendulum changes the period. This is the principle behind the workings of a grandfather's clock or any pendulum clock. There is usually a small threaded nut at the bottom of the bob that can be manipulated to raise or lower the bob, changing the period of the pendulum and therefore affecting the timekeeping mechanism of the clock. Lengthening the pendulum will slow it down and shortening it will speed it up.

The term *system* is an important scientific concept. A system is a group of related objects or parts that interact with each other. We have examples in our body: circulatory and respiratory systems; in astronomy: planetary and intergalactic systems; and we have examples in nature: soil and plant tissue systems and the larger *ecosystem*. One important aspect of a system is that all parts are interconnected so that a change in one part of the system has an effect on all other parts of the system. For instance in the Everglades, aquarium fish have been added, sometimes unwittingly, to the natural ecosystem and thus have changed the ecology. These fish—oscars, cyclids, tilapia, and other exotic species—have disrupted the food chain and the environment so that the native fish populations have suffered. Emphasizing this point of interconnectedness with your students is very helpful because it is a very important and all encompassing conceptual scheme that can increase understanding across the curriculum, even with nonscientific systems like politics and finance.

Systems are classified as open or closed. *Open systems*, such as the Earth, solutions, and ecosystems, allow substances to enter or leave the environments (which are probably systems in their own right). In a *closed system*, the materials that make up the system are enclosed in some way so that matter cannot enter or leave. An example of a closed system would be a nail in a moist environment enclosed in a container. Here, the only elements available to the objects in the system are those in the jar. You can get a better understanding by looking at two probes from *Uncovering Student Ideas in Science, Volume 4* (Keeley and Tugel 2009). "Is It a System?" will help students understand more of what constitutes a system and "Nails in a Jar" will show students how a closed system can be defined and used in a classroom as well as in developing a world view. In "Nails in a Jar," you are asked to predict whether, in a closed jar (closed system) rusting nails will cause the system to gain, lose, or conserve weight. In an open system, the nails would gain weight since elements from the outside would add substance to the nails, but in a closed system, only those elements in the jar can be involved in the chemical reaction. Therefore, the mass is conserved and there is no change in the weight of the system.

related ideas from the national science education standards (nrc 1996)

K–4: Abilities Necessary to Do Scientific Inquiry

- Ask a question about objects, organisms, and events in the environment.
- Plan and conduct a simple investigation.
- Employ simple equipment and tools to gather data and extend the senses.
- Use data to construct a reasonable explanation.
- Communicate investigations and explanations.

5–8: Abilities Necessary to Do Scientific Inquiry

- Identify questions that can be answered through scientific investigations.
- Design and conduct a scientific investigation
- Use appropriate tools and techniques to gather, analyze, and interpret data.
- Think critically and logically to make the relationships between evidence and explanations.

related ideas from benchmarks for science literacy (aaas 1993)

K–2: Scientific Inquiry

- People can often learn about things around them by just observing those things carefully, but sometimes they can learn more by doing something to the things and noting what happens.
- Describing things as accurately as possible is important in science because it enables people to compare their observations with those of others.
- When people give different descriptions of the same thing, it is usually a good idea to make some fresh observations instead of just arguing about who is right.

3–5: Scientific Inquiry
- Results of scientific investigations are seldom exactly the same, but if the differences are large, it is important to figure out why. One reason for following directions carefully and for keeping records of one's work is to provide information on what might have caused the differences.
- Scientists do not pay much attention to claims about how something they know about works unless the claims are backed up with evidence that can be confirmed with a logical argument.

6–8: Scientific Inquiry
- If more than one variable changes at the same time in an experiment, the outcome of the experiment may not be clearly attributable to any one of the variables. It may not always be possible to prevent outside variables from influencing the outcome of an investigation but collaboration among investigators can often lead to research designs that are able to deal with such situations.

USING THE STORIES WITH GRADES K–4

It is important that the children have an opportunity to look at the picture of the swing closely so that they can assess the parts of the system and note anything that does not look normal. If you can either draw the main parts of the swing system on the board or copy the picture, it will help them to analyze the problem. You may want to refer to the *National Science Education Standards* (1996, pp. 146–147) essay on pendulums to see how one teacher has a class explore pendulum systems. With elementary age children, it is probably best to have them explore simple pendulums first. Consider using the story "Grandfather's Clock" (Konicek-Moran 2008) mentioned above before pushing on to "The Crooked Swing."

With younger children, you will have to consider whether or not they possess the skills to count and keep records, although all they need to do is count the number of swings within a set time. Young children do play on swings that are pendulums. If these can be modified to shorten both of the supporting chains of one swing so that two swings next to each other can be compared, the idea of two different periods can be illustrated. Then the children can build their own pendulums and investigate their properties. (*Note:* tire swings are more simplified forms of pendulums since they are suspended from only one cord, therefore, they can never swing crooked.)

A wonderful article on helping students learn how to design inquiries is "Inquiry on Board," in *Science and Children* (Buttemer 2006). It is available online to members of NSTA. This article shows how variables can be identified and moved along a series of posters to form questions and finally to design an investigation. The crooked swing provides a wonderful opportunity for you to help your children look at a pen-

dulum as a system and to identify the separate parts. Asking them which parts could be changed without changing any of the others at the same time will begin to help them see how one goes about setting up an investigation.

My favorite way of setting up inquiry is to create a chart called, "Our Best Thinking Until Now," and ask the children to make statements about which of the variables might change how many times the pendulum will swing back and forth in one minute. Ask for reasons and write their predictions on the chart. Then change the statements to questions and explain that all of these questions can be answered by investigating them. Help them to design investigations that control the variables they are not testing. After testing the pendulums and recording their results in their science notebooks, you can discuss what they have found. Tell them they can go back to the chart and change any part of their statements that the data have shown need changing. This helps them realize that scientists can learn from predictions that are not correct as well as those that are, and that it is okay to change your mind when confronted with data that convinces you that you should.

If you decide to tackle the problem of the crooked swing with very young children, it is doubtful that they will focus their attention on the crooked branch and see this as the root of the problem. However, teaching is full of surprises! This can be demonstrated in the playground if the chains of the swing can be changed so that one side is longer than the other. But this results in a swing seat that is not horizontal, which may divert children from the point. The neat thing about the crooked swing problem presented above is that the swing itself remains horizontal just like any other porch swing that behaves itself. This fact stresses the importance of the lengths of the two support chains.

Upper elementary children are capable of making working models of the problem and should be encouraged to draw pictures of these models in their science notebooks. I always have materials on hand that can be used to make models. Scraps of wood, string, washers, and tape are usually enough for building working models at their desks. I find that working in groups on this problem is useful since several hands may be necessary to hold parts of the model in place while the testing goes on. Also, in this part of the problem-solving process, more minds at work mean more ideas to be tested.

USING THE STORIES WITH GRADES 5–8

Students in the middle grades are perfectly capable of looking at the swing picture, developing strategies to model the problem and find some solutions. They may not have had a chance to work with simple pendulums before and would benefit from working with some now before they try to apply pendulum theories to the new problem. They could follow the ideas suggested above and find out the variable that changes the period of the pendulum through direct experience. They may also need a review in setting up an investigation and controlling variables. It should not take more than one class period to prepare them to consider the prob-

lem of the crooked swing. As suggested above, they will benefit greatly by drawing the problem situation in their science notebooks and then working with a real model of the swing. They seem to work much better in groups both because they need each other's help in manipulating materials and because they need to discuss ideas in a group.

Some will immediately see that the problem lies with the fact that the branch angles upward, and will go on to suggest solutions. Most will have had experience with porch swings that were hung from horizontal ceilings, and will want to build a new structure on the limb. Some will suggest building a new free-standing structure on the ground from which to hang the swing. Others will find more creative ways of hanging the swing or finding ways to make the two suspending chains equal in length. Encourage this creativity by letting them brainstorm possible solutions. A colleague of mine taught me that when a student came up with a solution, I should answer in a positive way, such as, "Great idea! That's *one* way." It works. If you frame the problem in such a way that you ask for multiple solutions, you will find that students will not quit so easily. A reward for the most solutions might encourage them to look at the problem in several ways rather than in just one.

related NSTa Press Books and NSTa Journal articles

Keeley, P, 2005. *Science curriculum topic study: Bridging the gap between standards and practice.* Thousand Oaks, CA: Corwin Press.

Keeley, P., F. Eberle, and L. Farrin. 2005. *Uncovering student ideas in science, volume 1: 25 formative assessment probes.* Arlington, VA: NSTA Press.

Keeley, P., F. Eberle, and J. Tugel.2007. *Uncovering student ideas in science, volume 2: 25 more formative assessment probes.* Arlington, VA: NSTA Press.

Keeley, P., F. Eberle, and C. Dorsey. 2008. *Uncovering student ideas in science, volume 3: Another 25 formative assessment probes.* Arlington, VA: NSTA Press.

Keeley, P., and J. Tugel. 2009. *Uncovering student ideas in science, volume 4: 25 new formative assessment probes.* Arlington, VA: NSTA Press.

Konicek-Moran, R. 2008. *Everyday science mysteries: Stories for inquiry-based science teaching.* Arlington, VA: NSTA Press.

Konicek-Moran, R. 2009. *More everyday science mysteries: Stories for inquiry-based science teaching.* Arlington, VA: NSTA Press.

Konicek-Moran, R. 2010. *Even more everyday science mysteries: Stories for inquiry-based science teaching.* Arlington, VA: NSTA Press.

references

American Association for the Advancement of Science (AAAS).1993. *Benchmarks for science literacy.* New York: Oxford University Press.

Buttemer, H. 2006. Inquiry on board. *Science and Children* 34–39.

Keeley, P., and J. Tugel. 2009. *Uncovering student ideas in science, volume 4: 25 new formative assessment probes.* Arlington, VA: NSTA Press.

Konicek-Moran, R. 2008. *Everyday science mysteries: Stories for inquiry-based science teaching.* Arlington, VA: NSTA Press.

National Research Council (NRC). 1996. *National science education standards.* Washington, DC: National Academy Press.

CHAPTER 18
THE COOKIE DILEMMA

Mom was baking cookies—chocolate chip cookies—and Cerisa and her sister Barbara could hardly wait for the cookies to come out of the oven. At the moment they were only a promise, since Mom was busy mixing all of the ingredients. There was flour, baking soda, sugar, butter, eggs, and chocolate chips. Mom always mixed the dry ingredients first and then added the eggs and butter at the end. But the phone rang and it was for Mom. She left the bowl with the dry ingredients on the counter and talked to her friend for a while. It seemed like an hour to Cerisa and Barbara, but that is the way time passes when you are hungry for nice warm cookies!

When she came back to the cookie bowl she said, "Oh goodness, I lost my place and don't know if I put in the baking soda or not. If I didn't put it in or put in double the amount, it will probably ruin the cookies. I guess I'll have to start over again!"

She got out an identical bowl and began again. This time she made sure she added the baking soda first and then checked it off the recipe. After the dry ingredients were mixed, she was ready for the butter and eggs.

Wouldn't you know it, the front door bell rang and of course it was someone for Mom. While she was at the door, Dad came in and noticed the bowls on the counter. He thought, "I'll just push these bowls over here so I can use the counter space to make myself a sandwich."

When Mom came back, she looked at the bowls and said, "Did you move these?" to the girls.

"Nope, Dad moved them because he needed counter space," said Cerisa. "Which one is which?"

"Oh rats" said Mom, "I can't tell the difference! And I am not about to start over again! What a waste!"

Cerisa and Barbara both saw their visions of cookies going down the drain, but were not about to give up. Cerisa picked up the bowls and turned to her mom.

"Mom, we have been doing something in school called *Mystery Powders* and if you give us some time I think we can tell you what ingredients are in the bowls. I'll need some stuff from the cupboard, though."

Mom was, of course, delighted to have them solve this mystery and helped them find the materials they needed to test the contents of the bowls. And, within a few minutes they came up with the answer to the problem and now were able to get two batches of cookies when they had their answers.

"Amazing!" said Dad as he munched on his cookie a half hour later, "How did you do that?"

PURPOSE

This story is based a bit on the ESS (Elementary Science Study) activity "Mystery Powders." The problem with the original activity was that the students were introduced to the tests for various ingredients without a motivating hook. Alecia Peck wrote a similar story to motivate her third grade class in western Massachusetts. It was of course, finished with a cookie treat for all. The new modification then became part of the third-grade curriculum for the school district. There are two purposes: Have children engage in problem solving and use their collective knowledge to identify certain chemical compounds in a mystery mix.

RELATED CONCEPTS

- Chemical reactions
- Designing scientific investigations
- Problem solving

DON'T BE SURPRISED

Some of your students will not have had experience in testing for various food substances such as starch, baking soda, baking powder, sugar, salt, and flour. Some may not be aware of how to design an investigation and control for variables. During the discussion following the reading of the story, you will probably become aware of any deficits.

CONTENT BACKGROUND

You will mostly be testing for chemical reactions that occur when two or more substances are combined and react with one another to form new substances. These new substances have different properties than any of the original reactants alone. Thus, mixing vinegar and baking soda will result in a combination of the two with carbon dioxide as a by-product, which will appear as a fizzing gas. Since, neither of the two substances are gases, the appearance of a gas is a definite clue that a chemical reaction has taken place.

Changes can occur in matter in at least two different common and observable ways; physical changes and chemical changes. Changes in state, shape, or color are usually considered to be physical changes. Melting or freezing, molding into a different shape and becoming part of a mixture are all considered to be physical changes. Physical changes do not change the chemical properties of the substances involved. Evaporation, condensation, melting, freezing, and going from a solid to a gas (sublimation) are all examples of physical changes. Carving a piece of wood or hammering a piece of soft metal are also examples of physical change. Putting two substances together that mix but do not chemically combine is another

example. Mixing salt or sugar into water results in the sugar or salt dissolving in the water but the salt, the sugar, and the water are not changed in their chemical makeup. The water can evaporat, another physical change, leaving behind the sugar or salt as a residue, thus proving that the substance has not really combined with the substance in which it is dissolved. This and the others mentioned above are examples of mixtures, all of which encompass physical changes.

Chemical changes, on the other hand, do involve changes in the reactants at a molecular level and involve the formation of new compounds. Rusting is a common example of a chemical reaction that we have all witnessed. Oxygen reacts with iron to form a new compound, iron oxide, commonly called rust. Usually this happens in a moist environment and is aided by any additional elements in the moisture that create movement of charged atoms or molecules in the system. Charged atoms are called ions. These have either a positive or negative electrical charge due to an excess or deficiency of electrons. In a chemical reaction the product has different properties than any of the reactants involved. These chemical reactions either give off heat or light, or require heat or light to initiate them.

Two of the most common chemical reactions will be used in this story. First, iodine, in the form of potassium iodide, when added to starch (flour) will result in a blue or blue-black color change. Vinegar (acetic acid), when added to baking soda (sodium bicarbonate), will undergo a chemical reaction that results in the release of carbon dioxide gas, usually observed as fizzing or bubbling. Caramelizing sugar involves heating sugar until its temperature is above 100°C (210°F) which causes the oxidation or combination of oxygen with the sugar. The tantalizing odor you smell is the release of those gaseous compounds. Crème brule, that tasty French custard dessert, has a crusty sugar coating on top usually accomplished by adding sugar to the top of the custard and heating it with a portable torch. Finally, looking with the aid of a magnifying glass for the salt crystals in the form of little cubes usually identifies it easily. To be even more certain, salt will dissolve in water whereas flour will not. Baking soda will also dissolve in water, but will be the only one to react to vinegar as well. Sugar will dissolve in water, but will also respond to the fire test whereas salt will not.

related ideas from the national science education standards (nrc 1996)

K–4 Properties of Objects and Materials
- Objects have many observable properties including size, weight, shape, color, temperature, and the ability to react with other substances.

5–8 Properties of Objects and Materials

- Substances react chemically in characteristic ways with other substances to form new substances (compounds) with different characteristic properties. In chemical reactions, the total mass is conserved.

related ideas from Benchmarks for science literacy (aaas 1993)

K–2 Structure of Matter

- Objects can be described in terms of the materials they are made of (e.g., clay, cloth, paper) and their physical properties (e.g., color, size, shape, weight, texture, flexibility).
- Things can be done to materials to change some of their properties but not all materials respond the same way to what is done to them.

3–5 Structure of Matter

- When a new material is made by combining two or more materials, it has properties that are different from the original materials.
- No matter how parts of an object are assembled, the weight of the whole object made is always the same as the sum of the parts and when a thing is broken into parts, the parts have the same total weight as the original object.

6–8 Structure of Matter

- Because most elements tend to combine with others, few elements are found in their pure form.
- An especially important kind of reaction between substances involved the combination of oxygen with something else, as in burning or rusting.
- No matter how substances within a closed system interact with one another, or how they combine or break apart, the total mass of the system remains the same. The idea of atoms explains the conservation of matter. If the number of atoms stays the same no matter how they are arranged, then their mass stays the same.
- The idea of atoms explains chemical reactions. When substances interact to form new substances, the atoms that make up the molecules of the original substances combine in new ways.

USING THE STORIES WITH GRADES K-4

This story was first tested with third graders to fit within a unit on chemical and physical changes in matter. As mentioned before, the teachers were concerned that the students were merely following a specified worksheet and not seeing any relevance to their lives. So we wrote a story to which they could relate. After the story, they were shown two bowls of white powders and told that they contained the ingredients mentioned in the story. One bowl had all of the required dry ingredients, flour, sugar, baking soda, and salt. The other was missing the baking soda. They were promised a cookie party if they were able to tell their teacher what was contained in each of the bowls.

The children met with the teacher to discuss the problem and to reach consensus on a solution that would reward them with a batch of chocolate chip cookies. There seemed to be little confusion, and in about 35 minutes they had agreed upon a plan. They decided to test for everything in each bowl. They agreed that they would take a sample from each bowl and label it so that five teams of students would each receive a sample to test for the various materials.

When asked if they knew how to test for baking soda, many had had experience with making rockets or volcanoes using vinegar and baking soda and knew about the fizzing reaction. They remembered the smell of burning sugar. These children knew what salt looked like under magnification and knew that it would dissolve in water. They didn't know of a test for starch, so the teacher presented them with a demonstration of starch and iodine. She also demonstrated the safety factors involved in testing for each substance. All children wore safety glasses, and only did the burning tests with the help of the teacher. She set up a burner and supervised the placing of powders on an aluminum foil tray so that the students could observe the reaction with heat.

The next thing to do was to set up the experimental design for each test. The fact that the substances were mixed did not cause a problem since when putting the powders in a foil tube and heating, the flour gave off a burnt toast smell and the sugar gave off the distinctive burnt sugar smell, sweet and caramel-like. The students tried all tests and found that in the sample from one bowl, the vinegar test did not work but all other tests did. In the other bowl, all tests were positive. They concluded, therefore, that one bowl contained all the ingredients, and that the other lacked baking soda.

The one problem occurred when the teacher asked them how they would record their results. They did not see why it was necessary to form a table of ingredients and their response to each test. The teacher allowed each group to use their own method of keeping data. With some groups, there was no problem. With other groups they were pressed to remember the results of their tests. The teacher could have insisted that they use the table used in the ESS "Mystery Powders" booklet, but decided to let them see the need for a way to record data. She concluded that they did not see the problem until they were involved enough to

be aware of the need to revisit data. In subsequent situations, they were careful to work this out. It was only because of experiencing a problem that they were aware of a need.

The cookies were delicious and helped to conclude an experience that amazed the teachers who witnessed the process. Cookies were served while students had a group discussion to discuss the process and what the students had learned.

USING THE STORIES WITH GRADES 5-8

The process used with middle school children should not be too different than that with grades k–4. One teacher chose to add several different new tests that included testing for pH. Testing for acids and bases will give your students another way of categorizing matter. Acids are usually sour and bases are usually bitter, but taste is not a good way to test things. It can be quite dangerous! Instead we use indicators such as litmus or an easily made indicator extracted from red cabbage. I usually put some red cabbage in a blender with some water and blend thoroughly. I filter it through a coffee filter, producing a pinkish liquid. This liquid will change color when combined with an acid or base. It will become red with strong acids and turn greenish yellow with strong bases. Examples of acids found in a home are citrus juices, vinegar, and tomatoes. Soaps, detergents, hydrogen peroxide, and antacids are examples of bases.

What are acids and bases? Simplified, when acids or bases are in aqueous solutions, they break apart into (H+) ions and (OH-) ions. If there is a preponderance of (H+) ions in a solution, it is an acid and the opposite is true for bases. As you can probably guess, it they are put in a solution together, they come together and form water, a neutral substance. Thus, if you have an acid stomach and add an antacid (base), they are neutralized in your stomach and you get relief, at least temporarily. There will probably be a "burp" as your stomach gets rid of the gas.

The only chemical from the story that will give you a definite pH reading is the vinegar. However, receiving pH readings of neutral are still data. Since sugar and salt dissolve in water, you would not expect them to alter a pH reading. The same is true of cornstarch and flour. If your students get caught up in the use of the cabbage juice, you can help them to try detergents, antacids, citrus juices, and if you feel safe in demonstrating it, drain cleaning compounds, which, although highly toxic, give strong basic reactions. Students should never touch the latter, and of course safety glasses should be worn throughout the activities.

If you would like to have additional information and ideas on teaching about chemical and physical change, you can connect to NSTA's website (*www.nsta.org*) and members can download articles from past journals. Two interesting articles are "Using Easy Bake Ovens to Teach Chemistry" by Herald (2004), and "Enhancing Student Understanding of Physical and Chemical Changes," by McIntosh, et al (2009). I was excited particularly in how inexpensive and safe little ovens can be used in the classroom.

related nsta Press Books and nsta Journal articles

Keeley, P, 2005. *Science curriculum topic study: Bridging the gap between standards and practice*. Thousand Oaks, CA: Corwin Press.

Keeley, P., F. Eberle, and J. Tugel.2007. *Uncovering student ideas in science, volume 2: 25 more formative assessment probes*. Arlington, VA: NSTA Press.

Keeley, P., F. Eberle, and C. Dorsey. 2008. *Uncovering student ideas in science, volume 3: Another 25 formative assessment probes*. Arlington, VA: NSTA Press.

Keeley, P., and J. Tugel. 2009. *Uncovering student ideas in science, volume 4: 25 new formative assessment probes*. Arlington, VA: NSTA Press.

Konicek-Moran, R. 2008. *Everyday Science Mysteries: Stories for inquiry-based science teaching*. Arlington, VA: NSTA Press.

Konicek-Moran, R. 2009. *More everyday science mysteries: Stories for inquiry-based science teaching*. Arlington, VA: NSTA Press.

Konicek-Moran, R. 2010. *Even More Everyday Science Mysteries: Stories for inquiry-based science teaching*. Arlington, VA: NSTA Press.

references

American Association for the Advancement of Science (AAAS).1993. *Benchmarks for science literacy.* New York: Oxford University Press.

Elementary Science Study (ESS). 1957. *Mystery powders*. Newton, MA: Education Development Corporation.

Herald, C. 2004. Using Easy-Bake ovens to teach chemistry. *Science Scope* 27 (5): 24–29.

Keeley, P., and J. Tugel. 2009. *Uncovering student ideas in science, volume 4: 25 new formative assessment probes*. Arlington, VA: NSTA Press.

Konicek-Moran, R. 2008. *Everyday Science Mysteries: Stories for inquiry-based science teaching*. Arlington, VA: NSTA Press.

Konicek-Moran, R. 2009. *More everyday science mysteries: Stories for inquiry-based science teaching*. Arlington, VA: NSTA Press.

McIntosh, J., S. White, and R. Suter. 2009. Enhancing student understanding of physical and chemical changes. *Science Scope* 54–58.

National Research Council (NRC). 1996. *National science education standards*. Washington, DC: National Academy Press.

CHAPTER 19
STUCK!

Jimmy was looking forward to his afternoon at the playground. The parents in the neighborhood had just built a whole new set of equipment. There was a new slide that was higher than the old one and a lot faster. He had been there yesterday and had gone down really fast. It was a real thrill ride! He couldn't wait to get to the playground and try it again.

"Put on your shorts today, Jimmy," called his mother. "It is really hot out. Your jeans will be too warm to be comfortable."

"Okay, Mom," answered Jimmy.

He took off his jeans, put on his shorts and got ready to go. They had to take his new little sister and she had to be changed and it took a long time to collect all of her things. Babies seemed to be a lot of bother, but as his mother said, he was one too, once. But now he was a big boy ready to get to the playground and meet his friends from school.

There it was, shiny, metal, high, and slippery! Jimmy looked up at it and thought it was a good as any roller coaster because it curved down and gave him a real fast ride. Sometimes it was even a little scary, but he liked that too.

He climbed the stairs to the top and sat down on the downward slope of the slide with a whole line of kids behind him urging him to hurry up so they could have their turns. Jimmy let go and waited for the speed to begin on his ride to the bottom. Instead, he could hear squeaking on the slide as he went down slowly, and his legs underneath got really hot. He even had to push himself along in some places!

When he got to the bottom, he was very disappointed and ran to get in line to try again.

"Maybe they forgot to wax it or something," he thought as he climbed up the rungs. He finally got to the top and sat down again.

"Now!" he thought, "Let's fly!"

But the same thing happened. What was the matter? The only thing different from yesterday was that he was wearing shorts instead of jeans. Could that make a difference? All of the other kids were flying down even though they looked hot in their jeans.

PURPOSE

This story focuses on friction, both static and kinetic. Jimmy had trouble overcoming the *static friction* that needed a pushing force to get him started and then had trouble with his *kinetic friction*, his skin against the slide when he finally succeeded in getting moving with the help of the force of gravity.

RELATED CONCEPTS

- Friction
- Static friction
- Kinetic friction
- Net force
- Force
- Interaction

DON'T BE SURPRISED

Your students will probably not consider friction to be a force. It would be surprising if your students thought that friction was a force even in cases where there is no motion. They may also believe that there is no friction unless there is rubbing going on or that friction can occur with gases and liquids.

CONTENT BACKGROUND

Friction is a force, a fairly recent scientific theory. It is not one of the four *fundamental* forces: gravity, strong and weak nuclear forces, and electromagnetism. It is derived from the electromagnetic force between charged particles such as molecules, atoms, and the subatomic particles, electrons and protons. When two surfaces—be they solid, gas, liquid or any combination of these—interact, there is a conversion of kinetic energy into thermal energy or heat. The interacting surfaces are subject to chemical bonding and electrostatic forces. The roughness of the surfaces contributes to the amount of force, or *friction*.

Friction is often seen as the "enemy" of motion because it opposes it. The force of friction is always in the opposite direction of the force that acts on a body moving in one direction. This friction converts some kinetic motion into heat. It can cut down the efficiency of machines by losing this heat into the atmosphere. Physicists often work on problems where they use ideal situations, ignoring friction. In this way they can use formulas to calculate acceleration and other types of motion without including the muddying and not always predictable force of friction on the interacting bodies they are studying.

Friction can certainly be a problem, but we would experience a different kind of life if friction were not present in our everyday world. Without friction we

could not walk, run, drive, bicycle or even use our computer mouse. Can you imagine a world without it? On the positive side, you could slide to any destination in a straight line after having given yourself a push off. On the dangerous side, you would have a hard time changing directions and slowing or stopping when you came to an intersection or a wall, tree or even another person. We would all be like molecules bouncing off one another. Our cars and bicycles would not move because tires and roads could not push against each other. Our shoes wouldn't be able to push against the pavement, so we couldn't walk or run. Objects sitting on shelves would slide off at the slightest tremor. Anything in motion would remain that way and never stop. One exciting thing could happen—the illusive perpetual motion machine would finally be possible. That sounds wonderful, but the drawbacks, at least in my mind, outweigh the benefits!

Physicists talk about several types of friction, and here I will focus on two of them: *static friction* and sliding *kinetic friction*. When an object is placed on a flat surface, or at the top of an inclined surface, there are several forces acting upon it. First there is the force of *gravity* pulling down upon it toward the center of the Earth. We refer to this as weight. Second, there is the normal *force* that is the upward force that matches the downward force of its weight. This is Newton's Third Law that says for every force there is an opposite and equal force. Thirdly, there is the force of static friction between the interface of the object and the surface upon which it rests. This force resists motion just as does its inertia. *Inertia* is that property of matter that resists change in motion. Remember that according to Newton's first law of motion, an object keeps doing what it is doing until acted upon by an outside force. So the object is not about to move by itself. But when you do try to move it, there is not only the resistance of inertia, but there is also resistance because of the static (at rest) friction. We tend to assume that there is no friction unless there is motion. Scientists no longer believe this is true based on the current theory explaining that inter-molecular surface interaction is a cause of friction.

Once the object has enough force exerted upon it and it begins to move, overcoming both inertia and static friction, the friction of motion (kinetic friction) comes into play. The force needed to *keep* it moving is less than that which had to be applied to *get* it moving in the first place. This makes sense since now that it is in motion, the moving body has momentum and would require another force to stop it. This force is, of course, friction; not static friction, but kinetic friction. If you add the forward force caused by the push to the negative force of friction (opposing the movement), the *net force* would be the result. If the forward force is greater than the backward force, the net force would be in the direction of the positive force. In most cases, the net force *is* in favor of motion, so kinetic friction is less than the force of the push and momentum. In Jimmy's case, because he was wearing shorts and his skin provided more friction than the other kids' pants, the net force was in favor of friction, so he got stuck on the slide.

So we can see that friction is a force that acts between moving objects. We also can see that friction is our friend as well as our enemy. In the case of the story, Jimmy encounters both static and kinetic friction. At the top of the slide, he has trouble getting started because the static friction between his skin and the slide is too great for him to start moving easily. Then once he gets moving, his skin again rubs against the surface of the slide creating kinetic friction, and he is slowed down.

How can we measure friction? We often measure the resistance between two surfaces. These surfaces (say between the pavement and the rubber of the running shoe) are different, of course, and scientists assign each of these a value between 0 and 1, the higher number meaning a greater resistance. The ratio of these two numbers is called the *coefficient of friction.* Imagine that you are running on a grassy surface, and on one foot you are wearing a shoe with a slippery leather sole and on the other a rough, cleated running shoe. Which of the two would you think had the highest coefficient of friction? Go to any sports shoe department and examine the bottoms of running or sport shoes. They are designed to increase friction for obvious reasons.

We have all seen signs in public buildings telling us to use caution when floors have been washed. Moisture on tile floors reduces friction, and can be very slippery, causing loss of balance or worse. I have a smooth, slippery plastic seat on my lawn tractor and find myself in danger of falling off the seat when I turn corners. I have had to put a rubber mat on the seat for my own safety. That way my bottom can stick to the mat, and the mat sticks to the seat. A common item bought by those of us who travel in trailers is a plastic mat to put on shelves so that dishes and other utensils do not slide off the shelves while we drive from place to place. The mat has a rough surface on each side and is made of a material that forms a nice static connection to a shelf and to its contents.

While traveling through a construction zone the other day I was thinking about the new cement road blockers, their shape and material they are made of. They bulge out at the bottom and are more likely to come in contact with my tires should I be so unlucky as to scrape one. This would be to my advantage since my tires and the cement would provide a nice example of kinetic friction interaction, much more efficient and less damaging than if I hit a road guard made of metal at the level of my door. Now I appreciate that some engineer took friction into account when they were designed!

related Ideas From The national science education standards (nrc 1993)

K–4: *Position and Motion of Objects*
- An object's motion can be described by tracing and measuring its position over time.
- The position and motion of objects can be changed by pushing or pulling. The size of the change is related to the strength of the push or pull.

5–8: *Motion and Forces*
- Unbalanced forces will cause changes in the speed or direction of an object's motion.

related Ideas From Benchmarks For science Literacy (aaas 1993)

K–2: *Motion*
- The way to change how something is moving is to give it a push or a pull.

3–5: *Motion*
- Changes in speed or direction of motion are caused by forces.

6–8: *Motion*
- An unbalanced force acting on an object changes its speed, direction of motion, or both.

USING THE STORIES WITH GRADES K–4

Every child has encountered friction in one form or the other, although the name is not as evident as the effect it has had on his or her life. They can relate all kinds of pain and suffering visited upon them by that nasty force: carpet burns, mat burns in gym, skinned knees. If they have ever gone down a playground slide, this story will help them connect to the concept of friction.

You might ask them to list the friction episodes they have encountered, and then have them brainstorm ways in which something could have been changed so that those painful results might have been avoided or lessened. For example, if skinned knees

resulted from a game of tag on a sidewalk, things might have been different if they had played on grass. Grass has a smoother surface than concrete, so they would have slid and not scraped. Smooth mats can prevent mat burns; protective clothing and other safety precautions can stop scrapes, and so on. This can segue into the story and what Jimmy could do to avoid his disappointing ride down the slide. Let's look at a list of possible responses from the children on how Jimmy might slide faster:

- Jimmy should wear smooth pants.
- Jimmy should slide down on a towel.
- Jimmy should slide down on a sheet of newspaper.
- Jimmy should slide down on waxed paper.
- Jimmy could wet the slide.
- Jimmy could wax the slide.

From this list you can change each statement into a testable question. For example, would something slide faster if it were to be placed on a towel? Various materials can be tried on the slide to see which ones provide a faster trip down. A different tack would be to find materials that would cause a slower ride. If a slide is not available on your playground, you can use a slippery board on an incline for testing. Then use a board with rougher surfaces.

Older children might be able to use spring scales to pull objects of different frictional coefficients along a flat surface. I have used small blocks with various grades of sandpaper glued to one side and a cup hook for attaching the spring scale screwed into the end. Place a heavy weight on the block to make the difference significant enough to measure. Be sure to have the children pull in a horizontal direction as a controlled variable. We have found that it is best to use an old board on which to try this activity because heavy sandpaper can scratch up the surface of a desk rather badly. Have the children predict the effect of each material they use and give reasons for their predictions. If you ask them to find out if it takes more force to start a block moving than to keep it moving, you can discuss the ideas of static and kinetic friction with them.

A further extension for older children would be to find ways to eliminate as much friction as possible with oil, grease, wax, water, and other forms of lubricants. Explain that we are dealing with something called *sliding friction*, which is different than *rolling friction* where only the small bit of surface touching the incline is a factor. Since the coefficient of rolling friction is usually much smaller than sliding friction, rolling objects will react differently on an incline than sliding objects. Toy cars will zip down an incline while a sliding block will go much slower. This type of discussion is fairly high level, so you should decide if this is appropriate for your age group.

If you would like to explore more about motion with your students and look a little more closely at Newton's Laws, I recommend an article for younger children called Science Shorts: "Knowing Newton" in *Science and Children* (Ohana 2008). The author illustrates how using toy cars in experimental situations helps kids understand motion and friction.

USING THE STORIES WITH GRADES 5-8

Older students may be ready to discuss the empirical results of their investigations in the proper international units of Newtons. (Guess who they are named after?) Disregarding friction, a Newton is the amount of force necessary to accelerate a mass of one kilogram one meter per second. It is abbreviated as **N** (capitalized since it is named after a person). A kilogram is the international metric measure of weight. When we measure an object's *mass,* we are talking about the amount of matter and its resistance to change in motion. *Weight* is the gravitational mutual attraction between the object and the Earth. So there is a definite difference in the two speaking from a physics point of view. In the United States, outside of the scientific community, we commonly describe the amount of matter by its weight and use the term "pound." A Newton is much more appropriate than the kilogram when describing mass since it measures the inertia of an object and its resistance to change. One kilogram weighs 9.8 Newtons. If students want to do the conversion math, they will find out that one Newton is equal to about a quarter of a pound.

For many reasons, the United States has not joined the rest of the world in accepting the metric system of measurement. Instead, we find that when we communicate with people from other lands, we have to do numerical conversion. Some schools even teach their children how to convert from one to the other rather than having them think in metric. This is odd, since in science, the metric system is accepted worldwide, yes, even in America! For this reason, we are happy to see science classes here use the metric system and measure forces in Newtons.

The same activities described in the K–4 section can be done here as well. Students will want to test the static and kinetic friction of various surfaces and they can use spring scales calibrated in Newtons if they are available. You may also want to use metric weights on the blocks, since measuring is easier with heavier objects. The data should show that once the initial static friction is overcome and the blocks are moving, the kinetic friction is lower. Once the sandpaper is glued onto the blocks, it is easy to substitute different kinds of cloth since the sandpaper will hold the swatches in place for testing. Some teachers will refer to weighing things in metric units as "massing." I personally think that since mass is a property of matter, the word *mass* should be left as a noun and not used as a verb. However, you and your colleagues may want to argue this one out.

In addition to the above activity, using various grit levels of sandpaper will show why rougher sandpaper uses friction to take larger amounts of wood off of a board and why smaller grit is used for fine work. They can try using various kinds of sandpaper on wood and will be able to feel the heat and compare the differences.

To engage the students in an extension of the use of friction into areas other than pure physical science, I would recommend reading and using the ideas suggested in the *Science Scope* article "Tread Lightly: The Truth about Science Friction" (Chessin 2009). In this article, the author shows how animals have adapted to moving on different surfaces by having various ways of using friction. She also has a great activity using an assortment of sneakers and measuring their resistance to motion due to friction.

related NSTa Press Books and NSTa Journal articles

Keeley, P. 2005. *Science curriculum topic study: Bridging the gap between standards and practice.* Thousand Oaks, CA: Corwin Press.

Keeley, P., F. Eberle, and J. Tugel.2007. *Uncovering student ideas in science, volume 2: 25 more formative assessment probes.* Arlington, VA: NSTA Press.

Keeley, P., F. Eberle, and C. Dorsey. 2008. *Uncovering student ideas in science, volume 3: Another 25 formative assessment probes.* Arlington, VA: NSTA Press.

Keeley, P., and J. Tugel. 2009. *Uncovering student ideas in science, volume 4: 25 new formative assessment probes.* Arlington, VA: NSTA Press.

Konicek-Moran, R. 2008. *Everyday Science Mysteries: Stories for inquiry-based science teaching.* Arlington, VA: NSTA Press.

Konicek-Moran, R. 2009. *More everyday science mysteries: Stories for inquiry-based science teaching.* Arlington, VA: NSTA Press.

Konicek-Moran, R. 2010. *Even More Everyday Science Mysteries: Stories for inquiry-based science teaching.* Arlington, VA: NSTA Press.

references

American Association for the Advancement of Science (AAAS).1993. *Benchmarks for science literacy.* New York: Oxford University Press.

Chessin, D. 2009. Tread lightly: The truth about science friction. *Science Scope* 32 (6): 25–30.

Keeley, P., and J. Tugel. 2009. *Uncovering student ideas in science, volume 4: 25 new formative assessment probes.* Arlington, VA: NSTA Press.

National Research Council (NRC). 1996. *National science education standards.* Washington, DC: National Academy Press.

Ohana, C. 2008. Science Shorts: "Knowing Newton." *Science & Children* 45 (7): 64–66.

INDEX